T0317715

Arbitration Awards

A practical approach

Arbitration Awards

A practical approach

Ray Turner BSc, FRICS, FCIOB, FAE, FCIArb
Foreword by The Rt Hon The Lord Bingham
of Cornhill

Blackwell
Publishing

© 2005 by Blackwell Publishing Ltd

Editorial offices:
Blackwell Publishing Ltd, 9600 Garsington Road, Oxford OX4 2DQ, UK
  Tel: +44 (0)1865 776868
Blackwell Publishing Inc., 350 Main Street, Malden, MA 02148-5020, USA
  Tel: +1 781 388 8250
Blackwell Publishing Asia Pty Ltd, 550 Swanston Street, Carlton, Victoria
3053, Australia
  Tel: +61 (0)3 8359 1011

The right of the Author to be identified as the Author of this Work has been
asserted in accordance with the Copyright, Designs and Patents Act 1988.

All rights reserved. No part of this publication may be reproduced, stored in a
retrieval system, or transmitted, in any form or by any means, electronic,
mechanical, photocopying, recording or otherwise, except as permitted by the
UK Copyright, Designs and Patents Act 1988, without the prior permission of
the publisher.

First published 2005 by Blackwell Publishing Ltd

Library of Congress Cataloging-in-Publication Data

Turner, Ray (W. Ray)
  Arbitration awards : a practical approach/Ray Turner.— 1st ed.
    p. cm.
  Includes bibliographical references and index.
  ISBN-13: 978-1-4051-3063-9 (hardback : alk. paper)
  ISBN-10: 1-4051-3063-6 (hardback : alk. paper)

  1. Arbitration and award—England. 2. Arbitration and award—Wales. I.
Title.

KD7645.T825 2005
347.42′09—dc22

2004024236

A catalogue record for this title is available from the British Library

For further information on Blackwell Publishing, visit our website:
www.blackwellpublishing.com

Dedicated to

my wife, Beatrice, with love, and
with gratitude for her tolerance

and to the memory of
my late friends and fellow practitioners,
Leslie W.M. Alexander and Michael J. Needham
with both of whom I spent, if two decades apart,
so many fascinating hours planning and delivering
arbitration courses, or simply discussing arbitration

Contents

Foreword
by the Rt Hon The Lord Bingham
of Cornhill, Senior Law Lord

Ray Turner is the doyen of English lay arbitrators in the construction field.

Since becoming a member of the Chartered Institute of Arbitrators (in which he later became a leading figure) over 50 years ago, he has conducted several hundred arbitrations. In addition, he has lectured on arbitration for a number of universities and professional associations at home and abroad, becoming the first Visiting Professor of Arbitration at Leeds Metropolitan University. This book is the fruit of his great experience and expertise.

It is, as its title makes clear, a book about arbitration awards. There is, of course, more to being a successful arbitrator than the skill to prepare an award. But without that skill all other qualities of the good arbitrator – fairness, open-mindedness, authority, efficiency, clarity of thought – are devalued. For those who submit their differences to arbitration have the right to expect, whether they win or lose, that the process will culminate in a decision which is clear, coherent, comprehensive and such as can, if the need arises, be enforced by coercive legal process.

The arbitrator who has conscientiously considered the points argued before him (or her) and formed an objective judgement upon them will be well on the way to making a sound award. But even such a paragon will be helped by following (subject to any appropriate adaptation) a certain order or pattern of presentation; by adopting certain modes of expression; and by checking that everything which ought to have been covered has been covered.

The author of this valuable book wisely eschews any attempt to impose uniform and inflexible rules on individual arbitrators. But he does, drawing on his long and practical experience, give guidance which only the foolhardy would reject without good reason for doing so. With this manual beside him, many an arbitrator will, I feel sure, sleep the sounder.

Tom Bingham
House of Lords

Preface

Arbitration is an accepted means of finally resolving disputes in a wide range of areas of commercial and other activity, be it commodities or insurance, maritime matters or rent disputes, construction or commerce; the list is lengthy. Each area of activity tends to have its own requirements or traditions relating to awards, or to their style of presentation. Each arbitration can also have its own peculiarities which might demand a particular format or sequence for the contents of the award or awards.

Particularly as a consequence of the Arbitration Act 1996, arbitral procedures and the powers of the arbitrator can differ between cases, sometimes substantially so. The considerable party autonomy inherent in that Act can affect many aspects of any particular arbitration. It can and does affect the arbitrator's functions. Those functions might be those leading up to, and the conduct of, the hearing or its equivalent, or those involved in decision-making on the substantive and other issues, or the task of preparing an enforceable award incorporating those decisions.

The nature, format, terminology and sequence of an award depend upon the interaction of many separate permutations. They can include, but are certainly not limited to, the type of relationship between the parties, the nature of the dispute, any incorporated provision for arbitration, conditions of appointment, the type of remedy or remedies sought, the legal or other basis to be used for determining the issues, and then the overlaid effect of that party autonomy on the arbitrator's powers and the procedures to be adopted. The possibilities of inter-mixtures are near endless. The arbitrator's task of sieving and weighing the various legal and other submissions, of reaching a conclusion on the individual issues – and then drafting a valid enforceable award – can be substantial. Even in small and relatively straightforward arbitrations under the rules of some 'consumer' schemes, unusual problems can arise, not previously-encountered. In commercial disputes of substance, there can be a complex web of interlinked decisions to be made, then incorporated into an award, or a series of awards. Those awards might have to face the scrutiny of the courts.

The purpose of this book is not to suggest some kind of inflexible approach to the tasks, whether of decision-making or of award drafting. It outlines the nature of the decisions to be made and illustrates one possible approach to the preparation of awards; not the, or the only approach. There is no standard method of tackling the preparation for, and the writing of, an award, nor should there be. Two experienced arbitrators faced with the same submissions and evidence might well tackle both the decision-making process and the sequence and language of the award in different ways. Those differences are not important, provided that in every situation and with every arbitrator there is a rigorous and disciplined approach to decision-making, stemming from proper analysis of the submissions and evidence, and to the drafting of a valid and enforceable award which gives effect to those decisions.

Those facing the preparation of an award for the first time, be they mid-career examination candidates or arbitrators facing their first appointment, have no means of knowing whether that initial foray will be relatively straightforward, or be awash with the unexpected. Hopefully, so far as the practical tasks are concerned, this book will help better prepare them for what could be in store. It might also give support or solace, or suggest alternatives, to at least some existing practitioners – for a sole arbitrator's lot can be a lonely one.

This book is essentially a practical supplement to well-known works on the law and practice of arbitration. Although, perforce, these pages contain many references to the Arbitration Act 1996 and relate in substantial part to the legal background to arbitration, the citation of common law authorities is intentionally left for other reading. Part A is a preliminary introduction, mainly for the uninitiated. In Part B, types of award, style and content, then decision-making are considered. The example of a checklist illustrated there uses alphanumeric referencing to distinguish it from the main enumeration and to enable the narrative and illustrations in Parts C to E to be referenced back to it. Part C develops from that background and assembles an award. After, in Part D, expanding on the various identified types of award, other options and variations of content are explored in Part E. Other than in Part E, the illustrations intentionally adopt the same underlying background scenario, with the same parties, arbitral provisions and appointment. For the convenience of readers, the check-list illustrated in Chapter 3 and the award generated in Chapter 5 are both repeated, uninterrupted by explanatory narrative, as appendices in Part F.

It has been suggested to me that the contents, whilst intentionally referring to arbitration in general, i.e. in any area of activity, might also be of assistance to some of those people who act as construction industry adjudicators. That, it has been suggested, might be particularly so when related to some of the very substantial (and longer-term) adjudications which are apparently arising. Whilst not written with that purpose in mind I can see that, with appropriate adaptation to suit the specific legislation or scheme, the general concepts could be of assistance.

I express my sincere gratitude to fellow practitioners, particularly Douglas Craig and Michael Joyce, who have had the kindness and the fortitude to read over earlier drafts of this book. Any remaining flaws are entirely of my own making. I am also grateful to the team at Blackwell Publishing for their help and encouragement. I readily give credit, too, to those many parties and their advisers, and to countless students, all of whom have by posing their various disputes and questions made me think the wider.

Finally, I am immensely conscious of, and grateful for, the honour which Lord Bingham has done me by readily agreeing to write the Foreword to this book.

Ray Turner

Part A
Introduction

1   A preliminary introduction

1   A preliminary introduction

This introductory chapter is intended primarily for readers who are substantially new to the topic, but it might also be a reminder to others of the basic background to the subject. It and what follows in later chapters is intended to point the way, to provoke wider thought, to illustrate concepts, not to be advice as such – for every arbitration is unique. References to relevant law, including the Arbitration Act 1996, are intentionally brief, but the text in later chapters is cross-referenced to that Act. Wider reading and understanding of the applicable law is essential. It is assumed that the reader is familiar with the meaning of terms in general use in arbitration and in that Act.

As many arbitrations which the average reader is likely to encounter will be before a sole arbitrator, the term 'arbitrator' is used in this book rather than 'tribunal' to the extent that the context allows. The term 'arbitrator' is used in a gender-neutral sense – and where it is inappropriate or cumbersome to use that term, the convention is adopted of using the masculine personal pronoun to embrace both 'he' and 'she' – and similarly with 'him' and 'her', 'his' and 'hers'.

1.1   Purpose and nature

1.1.1 *Definition of an award*

An award is the decision of the arbitrator based upon the submission or submissions made to him in an arbitration. It can be made orally, but an oral award is not covered by the provisions of the Arbitration Act 1996 ('the 1996 Act') and oral awards are rare, or an exceptional ad hoc measure in conditions of urgency – followed by the same in writing. An award must be the consequence of an arbitrator deciding as between opposing contentions, having weighed the evidence and submissions.

1.1.2   Purpose of an award

The principal purpose of an award is to make, and to record, the arbitrator's final and binding decision on the matters in issue between the parties and, by publication of the award to the parties, to inform them of that decision. Every award must give reasons for the arbitrator's decision unless the parties agree that it shall not do so (or it is an agreed award confirming a settlement between the parties).

Other functions of an award are:

(a)   to facilitate an application by a successful party to the court for enforcement of the award in the event of failure by a losing party to comply with its provisions (provided that the award is 'in writing' as defined by section 5 of the 1996 Act); and

(b)   to enable an offended party to seek redress by challenging the award on the limited grounds provided by the 1996 Act.

1.1.3   Nature of an award

Depending upon what remedy has been sought by a claimant, an award might order a payment of money, or make a declaration as to any matter which has to be determined, or order a party to do or refrain from doing something, or order specific performance of a contract, or order rectification etc. of a deed or other document.

An award should, like a well-prepared report, succinctly set out the background and logically lead the reader through to the arbitrator's conclusion, how it was reached and what is ordered by it. In doing that, the award will amongst other things:

(a)   identify the parties and their relationship to each other (whether contractual or tortious), the fact of there being an arbitration agreement, the existence of a dispute within the scope of that arbitration agreement, the arbitral appointment, the substantive matters in dispute and the parties' submissions;

(b)   set out the arbitrator's consideration of the parties' submissions and evidence, together with his decision on the substantive matter(s), on interest if that is applicable, and on liability for costs incurred in connection with the arbitration; *and*

(c)   conclude with the arbitrator's unambiguous instructions or declaration.

This list paraphrases the content of an award and is intended to give a flavour to the complete newcomer to the topic before launching him into a brief introduction to the requirements of an award and then into a more detailed exploration of the topic. That exploration is intended as a precursor to, or reminder of, more detailed reading elsewhere.

1.1.4 Requirement for natural justice

In making an award, as with all his arbitral duties and in common with other tribunals, an arbitrator must act (and be seen to act) in accordance with the principles of natural justice, i.e. must:

(a) act fairly, in good faith and without bias;
(b) give each party a fair opportunity to present its case, to be aware of its opponent's case and to contradict that case; not to hear one party behind the back of the other party; *and*
(c) not have any interest in the subject-matter of the dispute.

1.2 Relevant law

1.2.1 General

Before considering the practical tasks involved in the preparation of and the essential content of an award, one should be aware of the nature and sources of the law which governs such awards. The illustrations in later chapters are, with slight exception, based upon English law, particularly as related to the validity, content and form of an award. The reader of this practical manual should already be reasonably familiar with, or at least have read, the law governing the topic, including the 1996 Act in its entirety.

Arbitration is controlled by a combination of public law (under English law, and in this context, being a combination of statute law and common law) and private law.

Statute law is that laid down by Parliament. The Arbitration Act 1996 relates solely to the conduct of arbitrations, including the form, content and effect of awards. Other statutes can come into play in the decision-making process, including for example the Civil Evidence Act, the Unfair Contract Terms Act, and so on – but they are for study elsewhere.

Common law is the body of law created and developed by the courts, as recorded in the various law reports. It is the subject of continuing refinement, amendment and refreshment.

Private law consists of the legally binding agreements between the parties – e.g. any underlying contract, the arbitration agreement, any applicable rules, and so on.

The 1996 Act incorporates elements of private law by considerable provision for party agreement on procedures and other matters, including the form of awards. Additionally, by section 81, past common law authorities remain intact (and therefore applicable) provided that they are consistent with Part 1 of the 1996 Act.

In these increasingly global times, whilst it will be obvious to a newcomer to this subject that in a contractual dispute with international implications, *the proper law of the contract* will impinge upon the decision-making process prior to the drafting of an award, it will be less obvious to such newcomer that there can also be involved:

(a) the law governing the obligations of the parties to submit to arbitration – i.e. the proper law of the arbitration agreement; *and*
(b) the law governing the conduct of the particular reference – commonly termed the 'curial law';

any of which proper laws might be a law other than English law. That applicable law might be a specified law agreed by the parties to apply, or it might be one of the matters in contention in the arbitration.

1.2.2 The Arbitration Act 1996

Many topics arising throughout the following chapters are cross-referenced to the appropriate section(s) of the 1996 Act. That is in addition to the limited, and intentionally brief, references to specifically award-related sections of the Act which are shown below. To expand on them here, however briefly, or to include other potentially award-relevant sections, could be to bait a trap for the newcomer to the subject by deflecting him from more comprehensive reading. The whole Act, and more, should be studied in detail.

Some of the provisions of the 1996 Act are positive and direct in their applicability to awards. Others are indirect but can well influence not only the format of an award or what might need to be recited in it, but also the basis of determination of the substantive issue(s).

Provisions of the 1996 Act arise:

(a) as mandatory provisions, being those referred to in section 4(1) as listed in Schedule 1 of the Act; or
(b) by party agreement ('party autonomy') where the various sections so allow; *and*
(c) as 'default' provisions, which come into play unless the parties agree otherwise.

These are identified in the lists below by 'M', 'PA' or 'D', respectively. The party autonomy provisions are phrased in several different ways, but this is not the place to explore the significance.

Some relevant sections are shown in the Act under the specific heading of *Awards*. These are:

section 46	PA/D	Rules applicable to substance of dispute
section 47	PA/D	Awards on different issues &c
section 48	PA/D	Remedies
section 49	PA/D	Interest
section 50	PA/D	Extension of time for making award

section 51	PA/D	Settlement
section 52	PA/D	Form of award
section 53	PA/D	Place where award treated as made
section 54	PA/D	Date of award
section 55	PA/D	Notification of award
section 56	M	Power to withhold award in case of non-payment
section 57	PA/D	Correction of award or additional award
section 58	PA/D	Effect of award

Further mandatory sections which can have a direct or indirect impact upon the preparation of an award include but are not restricted to:

section 31	M	Objection to substantive jurisdiction of tribunal
section 32	M	Determination of preliminary point of jurisdiction
section 33	M	General duty of the tribunal
section 37(2)	M	Fees and expenses of expert, etc., appointed by the tribunal
section 60	M	Agreement to pay costs in any event

Other (non-mandatory) sections also have a direct bearing on the content of an award, including:

section 30	PA/D	Competence of tribunal to rule on its own jurisdiction (see also the related mandatory section 31 (above)
section 41(3) and (6)	PA/D	Award dismissing a claim following the faults described
section 45(1)	PA/D	Effect of agreement to dispense with reasons
section 57	PA/D	Correction of award or additional award
sections 61–65	PA/D	Costs of the arbitration
section 69(1)	PA/D	Effect of agreement to dispense with reasons

See Chapter 2, at 2.6, and Chapter 8, at 8.4, regarding section 39, Power to make a provisional award. In usual circumstances, such provision is considered to be by way of an *Order* rather than an award).

So far as the content of an award is concerned, section 52 lays down default requirements whereby the award shall:

(a) be in writing;
(b) be signed (see also under 'Formal requirements' later);
(c) contain the reasons for the award unless it is an agreed award or the parties have agreed to dispense with reasons;
(d) state the seat of the arbitration (i.e. the 'juridicial seat' – see section 3 for definition and section 2 and 53 for relevance); *and*
(e) state the date when made (and see section 54 on what is taken as to be the date on which made).

Relevant sections are referred to again below under 'Requirements', where it is the default provisions which are shown. There are also cross-references, in the check-list and illustrations in later chapters, to relevant sections of the 1996 Act. It is, however, important to realise that the impact of party autonomy, or provisions contained in the arbitrator's terms of appointment, can mean in a particular arbitration that some of the 'default' provisions do not apply, or are amended. Thus the procedures and arbitral powers can vary from arbitration to arbitration, with a consequent effect upon the nature and content of the award.

1.3   Requirements of an enforceable award

To be enforceable an award must be a valid award and be of a nature which is capable of enforcement (see below). The requirements of an enforceable award stem from a combination of statute, common law, party autonomy (i.e. provisions of the arbitration agreement, or subsequent agreement) and any applicable rules.

Various commentators identify a range of apparently different sets of requirements, but in the main these are simply different ways of expressing the same fundamentals. They are often classified as 'formal' and 'substantive' requirements, although different commentators list these in varying ways – and for purposes of drafting an award there is little, if any, need to distinguish between the two.

1.3.1   *Formal requirements*

These requirements might be said to stem from two sources:

(a) those now required by or under the provisions of the 1996 Act (including the provisions of any applicable rules); *and*
(b) those which are desirable as a matter of common practice so as to facilitate enforcement, as well as to facilitate better understanding of the award by a reader.

These requirements (which again must be complied with by the arbitrator unless agreed otherwise by the parties where the 1996 Act so allows) are, in the main, as follows.

1.3.1.1   Writing and signature

To come within the ambit of the 1996 Act the award must (by section 52(3)), unless otherwise agreed by the parties, be 'in writing' and must be signed 'by all the arbitrators or all those assenting to the award'. (The use of the plural is to embrace tribunals of more than one.) Oral awards are not dealt with in any detail in this book, for it is better to think of such awards very much as a rarity and they are 'awards' to which the 1996 Act does not apply.

Whilst witnessing of the signature(s) is not strictly necessary, it is a prudent step. There can clearly be occasions (for instance the death of an arbitrator after making an award but before notifying the parties, but there can be others) when doubt might be cast upon whether or not what purports to be the arbitrator's signature is in fact so. If witnessed, the witness can testify to that fact.

1.3.1.2   Identification of the parties

The parties should be properly named initially and not simply referred to by general terms such as 'the owner', 'the tenant', 'the supplier', 'the purchaser' and so on. It is usual to name and identify the parties and their status ('claimant' and 'respondent'), but there is no reason why one of those other terms such as 'the tenant' could not thereafter be used once the party had been properly identified and such term defined (e.g. 'The claimant is Thomas Smith, the tenant of the premises in question ("the tenant")'). That would in general circumstances only be done where such term helped in the understanding of the award.

1.3.1.3   Recitals

Although the traditional term 'Recitals' is still used in some leading textbooks, many arbitrators now use words such as 'Background' or 'Preambles'. By whichever name, they constitute the recital (i.e. the setting down) of the background to the case – including matters relating to identification and jurisdiction, to procedural matters, and so on.

A common error is to make recitals far too comprehensive. By so doing not only is such an award likely to be unnecessarily prolix, it can also cause problems – for instance by generating anomalies or apparent conflicts, or possibly a lack of consistency. Something wrongly recited can at least cast doubt upon the quality of the decision itself – or even make the award unenforceable.

1.3.1.4   Reasons

Where the parties have not agreed otherwise, an award must (by section 52(4)) contain the reasons for the award. What is *not* required is a blow-by-blow account, nor every element of reasoning. An agreed award is still 'an award of the arbitrator [tribunal]', and must be so stated. The process is explored in Chapter 4 and illustrated in Chapter 5 onwards. Reference has already been made to sections 45(1) and 69(1), whereby an agreement to dispense with reasons acts as a bar to the court's jurisdiction under those sections, including that of appeal on a point of law.

1.3.1.5 Date

Unless the parties agree otherwise, the 1996 Act (by sections 52(5) and 54) requires an award to state the date when it is made. There are very practical reasons why an award should usually be dated, some of them being (unless agreed otherwise by the parties, where that facility exists in the provisions of the Act):

(a) If an award is a monetary award the date establishes (directly, or indirectly by reference to that date) the commencing date for any post-award interest.
(b) It establishes the date from which that particular award becomes final and binding. As indicated in Chapter 2, Section 2.3.3, there could be a series of awards in a reference.
(c) It fixes the date when, without delay, the award must be notified to the parties (under the Act, that is by service).
(d) It establishes the commencing date of the period within which various actions in connection with an award can be made, such as correction of an award, issue of an additional award dealing with a matter presented to the arbitrator but not dealt with in the award, the making of an application for such correction or additional award – and similarly with application or appeal under sections 67, 68 or 69 of the 1996 Act.
(e) It fixes the date at which the arbitrator becomes what is termed *functus officio* – i.e. the arbitrator's powers and functions have ceased in respect of the matters determined by that award, other than functions which are specifically empowered under the Act (such as to issue a corrective award).

1.3.1.6 Statement of seat

Unless otherwise agreed by the parties, the award shall (by section 52(5)) state the seat (i.e. the 'juridical seat' – see sections 2 and 3) of the arbitration – of importance where there are international implications. The designation of seat establishes which law is the applicable procedural law.

1.3.1.7 Issues dealt with

Where there is, or is to be, more than one award at different times on different aspects of the matters to be determined, each such award must identify the issue(s), or claim(s), or part(s) of the claim, dealt with in that award. (Also see under 'Completeness' below.)

1.3.1.8   Notification

There would be little point in an arbitrator making an award if he told no-one of its existence and its contents. It must therefore be published. Unless otherwise agreed by the parties the award must (by section 55(2)) be notified to them, i.e. all parties, by service of copies, which must be done without delay (subject to the provisions of section 56.)

1.3.2   *Substantive requirements*

A well-recognised list of the substantive requirements for a valid award is set out at 1.3.2.1 to 1.3.2.4 below. Requirements which are cited by some commentators as being further 'substantive' requirements, whilst others consider them as 'formal' requirements, or as being already embraced by those set out above, follow at 1.3.2.5 to 1.3.2.10.

1.3.2.1   Cogency

An award must be based upon, and show, convincing, persuasive and consistent reasoning.

1.3.2.2   Completeness

An award must deal with all matters with which it purports to deal – all matters in issue; no more than those.

A single dispute can embrace a number of issues. That is commonly the case. Conversely, there can be more than one award relating to a single arbitration, each award dealing with one or more separable issues. However, each such separate award must resolve all matters with which it purports to deal. Taken collectively, the separate awards must deal with and determine each and every issue in that arbitration.

1.3.2.3   Certainty

An award must not in any way be ambiguous. It must leave no doubts as to the intention of the arbitrator or what is to be done by one or more parties.

By way of example (but not here setting out the full wording), in a monetary award the operative part (i.e. the 'instruction') should say something like: 'I award and direct that the Respondent shall pay to the Claimant...', *never* something like 'I have formed the opinion that the Respondent ought to...'. That latter type of wording has no place in an award.

1.3.2.4   Finality

An award must not leave an opportunity for reopening the issues resolved by that award. It must be unconditional. Other than matters specifically reserved to a further award (e.g. costs), nothing must be left undecided.

All issues to be resolved by the arbitration must be determined by the arbitrator. To leave any to be resolved by a third party is not final and such an award would not be valid. The arbitrator cannot delegate a judicial function.

1.3.2.5 Enforceability
Section 66 relates to enforcement.

An award must be in a form capable of being enforced. For instance, if it awards a payment of money it must be clear as to just what amount is awarded, to be paid by whom to whom, when, and in what currency. If it orders performance of a contract it must precisely specify what is to be performed and the time by which it is to be performed.

An award making a declaration cannot, of itself and directly, be enforced by the court. However, subject to it complying with the other requirements, it is nonetheless a valid award upon which other actions could be founded if breached. The various types of award are discussed later.

1.3.2.6 Jurisdiction
Clearly an award can include only matters which are within that arbitrator's jurisdiction in respect of that arbitration (see sections 30 and 31 – and 66(3)).

1.3.2.7 Legality
An arbitrator cannot order a party to commit an illegal act.

1.3.2.8 Possibility
That which is awarded must be capable of performance.

1.3.2.9 Consistency
There must be no inconsistency in the award. For instance, if incorporated reasons clearly and incontrovertibly lead towards a decision for a party, yet the award finds against that party, the award would be unenforceable in that form, i.e. without correction.

1.3.2.10 Compliance with submission
The issues which are in dispute and are submitted to the arbitrator to resolve by an award constitute the submission. That submission might, by agreement of the parties, incorporate requirements relating to the award or awards – such as format, content, timing and so on. Such agreement might also be by the agreed adoption of applicable rules. The award must comply with those requirements, indeed with all aspects of the submission.

In preparing an award an arbitrator must have proper regard to and comply with all of the matters listed above. It is stressed that this book is aimed at practical aspects of award preparation. Adequate comprehensive awareness of the requirements of an award – and other aspects of arbitration – is dependent upon wider reading.

1.4   Distinctions from a judgment

For those new to the process of preparing an award and nervous of the prospect, much can be gained by studying the judgments of respected judges, provided that the essential differences between litigation and arbitration are borne in mind. There are distinctions which influence content and thus format.

One obvious distinction is the nature of the tribunal. In the case of litigation the tribunal is made up of one judge (or more if on appeal). In the case of arbitration it is a person (or several if a tribunal of more than one) appointed by the disputants in that particular reference, or by someone effectively acting on their behalf under the terms of an arbitration agreement (or the terms of a statute which provides for disputes to be resolved by arbitration).

The position of a judge is entirely one of status, with no contractual element between him and the parties. The position of an arbitrator, in marked contrast, varies from case to case with different proportions of status (derived from statute and common law) and contract or quasi-contract (derived principally from that arbitrator's terms of appointment on that arbitration). Those terms commonly differ from arbitrator to arbitrator, and with any one arbitrator might vary from case to case. The consequence of that distinction between judges and arbitrators is that whilst a judge has no need in a judgment to indicate how he comes to be there and by what authority, an arbitrator in an award needs to give or adequately indicate the existence of that information, for it underwrites and gives the source of authority for that which follows in the award.

In litigation the parties are well defined within the overall court procedure, so that little by way of overall identification needs to appear, and seldom does appear, in a judgment. By contrast, an award is not part of that system unless and until it is referred to the court for enforcement, or on appeal or for some other purpose. Particularly where one or more party is an individual or a firm with no registered address, the award needs sufficiently to identify the parties from other people having the same or similar name.

Another distinction relates to procedures and powers. The procedural powers of a judge are the same as another judge at that same level in the judicial hierarchy. The powers and functions of an arbitrator can, and do, vary from arbitration to arbitration. Such matters in arbitration are affected, to the extent allowed under the Arbitration Act 1996, by party agreement, by any applicable rules, and by that arbitrator's terms agreed with the parties. In the event of application for enforcement or for appeal, the exercise of some possible but perhaps unusual powers, if not referred to in the award along with their source, can create a situation or arrive at an award which could be inexplicable to the court unless it were informed of those powers and their source.

1.5 Illustrative monetary award

Readers who are new to the topic might not have seen an award. They are invited to look at the example in Appendix 3, as a foretaste of Chapters 2–5. That illustration should not, however, be taken as an exemplar. Some arbitrators would approach differently the matters of format, sequence and the extent of reasons included. As indicated in Chapter 3, there is no mandatory style for an award.

Part B
Background principles

2 Types of awards

2.1 Introduction

To put the following chapters into context for readers who are relative newcomers to the subject, it is appropriate first to consider the various types of award which can arise. The function of each type is identified at the beginning of each description, explaining the operative part of such an award, i.e. that which expresses the instructions, or declaration, of the arbitrator.

Types of awards are briefly explained in this chapter, and further illustrated in Chapters 6–9, under the following heads, being terms coined solely for ease of cross-reference in this book:

- 'substantive' awards, i.e. awards ordering one or more of the remedies listed in section 48 of the 1996 Act;
- 'supportive' awards, being those which have a close link with, or help to complete, those substantive awards;
- 'procedural' awards, here being those effectively terminating the reference;
- 'institutional' awards, being those under rules, other statutes or 'consumer' schemes;
- 'ancillary' awards, which are of an interlocutory nature.

The various remedies can only be awarded if the parties have not agreed otherwise and if that remedy has been claimed. It is not open for an arbitrator, of his own volition, to make a different type of award from that which was sought. He cannot, for instance, make a performance award when the claim was for money, with no alternative being claimed – unless the claimant has applied and has been permitted to amend its pleadings. Regarding performance, injunctive and rectificative awards, the power of the arbitrator, by section 48(5) of the 1996 Act, is 'the same power as the court'. Arbitrators should have clear regard to the manner in which the court acts in these matters, but that must be tempered by compliance with section 33(1)(a). By that provision the arbitrator must not only act fairly and impartially as between the parties, but must also give both parties a reasonable opportunity of putting his case and dealing with his opponent's case.

Comments and illustrations in this and later chapters are on the basis that all 'default' provisions of the 1996 Act apply without alteration and that there are only two parties. If any of the default provisions are altered by party agreement, the narrative and illustrations shown need suitable amendment.

2.2 'Substantive' awards

An award which gives a remedy in the substantive dispute might by section 48 be one or a combination of:

- a monetary award;
- a declaratory award;
- a performance award;
- an injunctive award;
- a rectificative award (here, and later, a shorthand term for an order of rectification, or setting aside, or cancellation.)

For convenience, these are listed in a sequence which reflects the general likelihood of them being met by an arbitrator.

It is possible that in a multi-issue claim, or a claim and counterclaim situation, either:

(a)   the same kind of remedy will be claimed under all issues; *or*
(b)   different remedies might be claimed under the various issues.

In either situation, the particular circumstances will determine whether all are included in a single award or a series of separate awards.

It is commonplace for a claim for (and the defence against) any of these remedies to include a claim for costs – that is the recoverable costs of the arbitration. The arbitrator's decision on that element can form part of the award which deals with the substantive matter, or can be reserved (i.e. postponed) to a subsequent award. For the purposes of this chapter, the latter has been assumed. To avoid confusion when indicating the nature of the content of the operative parts of the various types of award, no reference is made in this chapter to costs. However, in practice costs do figure in the operative parts of many awards. The topic of costs as such is outside the parameters of this book – other than how decisions on costs might be made and incorporated in an award.

It would be potentially confusing at this stage to develop the possibilities of mixed awards, or of such things as a monetary award following default in compliance with a performance award. Such matters are more appropriate later in the book.

2.2.1   *Monetary awards* (section 48(4))

By the operative part of a monetary award the arbitrator orders the payment of money by one party to another party.

The most common type of claim in arbitration is one for a payment of money, i.e. seeking a monetary award ordering a party to pay money to the other party, e.g.:

(a)   for work done or services provided; *or*
(b)   as damages for some breach of contract or in respect of a tortious act.

Such award might be for payment of:

(a)   an amount of money; *or*
(b)   an amount of money plus interest; *or*
(c)   interest alone (but only if interest had been reserved from an earlier monetary or declaratory award, or if other circumstances warrant it, e.g. under section 49(3)(b) of the 1996 Act).

A sum awarded can, by virtue of section 48(4) of the 1996 Act, be ordered to be paid in any currency, but adequate grounds would need to have been shown for payment other than that used in, for instance, the relevant contract.

2.2.2   *Declaratory awards* (section 48(3))

By the operative part of a declaratory award the arbitrator makes a declaration resolving a particular issue.

Situations where a declaratory award would be issued include, by way of example, disputes on:

(a)   the meaning or interpretation of a term of a contract, or specification, or other document; *or*
(b)   whether or not goods or services supplied comply with the terms of the contract of sale or its equivalent, or are 'fit for purpose';
(c)   whether or not the terms of a lease had been complied with;
(d)   whether at the time in question a vessel was an 'arrived ship'; *or*
(e)   whether one event was the direct consequence of another (for instance in an insurance dispute the question of whether the collapse of a temporary grandstand was in consequence of extreme weather conditions or of inadequate design).

Another likely situation is where there is a dispute as to whether or not a party is entitled to money from another party, but there is no dispute as to the amount which would be payable if liability were to be determined – or the parties believe that they could readily settle what that amount should be – or will proceed with submissions on the monetary consequences once the matter of liability has been established. In such a case the claimant would seek a declaratory award solely on liability.

Where the consequence of a declaratory award is that a party owes money to another party, that declaration can also cover interest (see section 49(5) of the 1996 Act.)

A declaratory award might be that, for example:

(a) the [respondent] is [or is not] liable in respect of the matter alleged;

(b) the claimant's [or the respondent's] interpretation of the clause in contention is correct; *or*

(c) that the relevant term(s) of the lease had [or had not] been complied with.

Whilst not directly enforceable in the same manner as an award which imposes an obligation, a declaratory award can be relied on in subsequent legal proceedings, provided it is of a nature which can be recognised (in the sense of 'accepted') by the court.

2.2.3 *Performance awards* (section 48(5)(b))

By the operative part of a performance award the arbitrator orders a party to perform some contractual obligation ('specific performance') in respect of which it has been found to be in default. (This power is not applicable to a contract relating to land.)

Some disputes arise as a consequence of a party alleging that the other party has failed to comply with, or possibly to complete, an alleged contractual obligation. An example could be where there is a contract for the design and installation of computer software together with the production of a user's manual and the training of specified staff in its use. The contract has been completed except for provision of the manual and the training. The provider has failed, or refused, to complete those parts of the service. The customer could seek a performance award, ordering the provider to complete that which it had undertaken to do under the contract.

A caveat is appropriate here. In some instances the performance ordered and purportedly executed might turn out to be, or be alleged to be, still non-compliant with the contract or award. It might be claimed by the party entitled to it that there has been no attempt at such performance. It is not for the arbitrator to supervise its execution.

It is desirable, when a performance award is sought, that the parties agree from the outset that in the event of failure to comply fully with a performance award, the offended party can make submissions for, and the arbitrator is empowered to make, a monetary award for damages. Such a situation is illustrated at 6.6.

2.2.4 *Injunctive awards* (section 48(5)(a))

By the operative part of an injunctive award the arbitrator orders a party to do or to refrain from doing some specified thing.

An injunctive award might, for instance:

(a) order a party to remove an obstruction to the claimant's window; *or*

(b) order a party to refrain from polluting another party's waterway.

The point has in effect already been made at 2.1, by reference to section 33 of the 1996 Act, that in considering and making, or refusing to make, an injunctive award the arbitrator must give each party 'a reasonable opportunity of putting his case and dealing with that of his opponent'. To that extent, an arbitrator could need to exercise more caution than might the court in similar circumstances.

2.2.5 *Rectificative awards* (section 48(5)(c))

By the operative part of what is here termed a rectificative award the arbitrator orders that a deed or document shall be rectified, set aside, or cancelled, as the case might be.

Experienced arbitrators will treat the preparation of such an award with some circumspection. For inexperienced arbitrators it might well be a potential minefield.

2.2.5.1 Rectification

Rectification is 'to put right, to amend'. In present-day commercial negotiations there can arise a welter of documentation, proposals, counter-proposals, offers and counter-offers, and so on. These can lead to mutual mistake in the final drafting of a resultant agreement, be it a contract, a lease, or other document. Such mistake might not be recognised until long after the contract has been executed, as a deed or otherwise. The subject matter of the agreement can be well under way, or even completed, when such alleged mistake is noticed. At that time it is likely that one party would wish the document to stay as it is – and be binding as it is – whilst the other party would wish to see it altered to what it thought it had initially agreed.

That is one of the situations, given an arbitration agreement which embraces that type of dispute, when the party wishing to have the deed or other document 'corrected' might seek an award rectifying that document – a rectificative award. The grounds for so applying for rectification are, it might be alleged, that mutual mistake or a drafting error occurred whereby the document did not reflect the true intention of the parties.

2.2.5.2   Setting aside

Setting aside is 'to reject, disregard, annul'. If a deed or other document is set aside, it has no further effect, at least in respect of the matter in dispute.

It might be that a party has undertaken, perhaps but not necessarily ancillary to some other agreement, to refrain from some action or another. It then alleges that it should not be bound by that undertaking because it was procured by serious misrepresentation or by fraud. It cannot agree with the other party. Provided that the dispute is covered by the terms of the arbitration agreement, such party can apply to the arbitrator to set aside that undertaking.

In some instances it might well be that there is little or no difference in effect between rectification and setting aside. Such an instance might be where a document or deed has some other document annexed; one party contends that it should simply not be there, as having no part in the purpose of the deed or other document, but the other party disagrees. Setting aside that allegedly wrongly annexed document would have substantially the same practical effect as rectifying the deed or document by excluding that annexure.

2.2.5.3   Cancellation

Cancellation is 'to disregard, to annul' and deprives a document of its effect. Once again, whatever the legal distinction between the terms, there could be situations where there would be little practical difference between setting aside and cancellation.

An example of an application for cancellation of a document might be where a 'letter of intent' had been issued by one party to another, undertaking to enter into a subsequent contract. The issuing party later alleges that in the prevailing circumstances, it should no longer be bound by that document. The other party disagrees and, again provided that the arbitration agreement was in terms which embraced that type of dispute, the party seeking to be relieved of the obligation can seek an award cancelling that letter.

2.3   'Supportive' awards

The 1996 Act makes specific provision for a variety of matters to be determined by an award. Some deal with remedies, others do not. These, and several related matters, are set out below, as follows:

- an award on jurisdiction;
- an agreed award;
- an award on a separate issue;
- an award on a reserved matter;
- a corrective or an additional award;
- an award following remission;
- an award giving further reasons;
- an award following payment during the reference;
- an award following:
  - an earlier order giving provisional relief under section 39, *or*
  - an earlier adjudicator's decision (e.g. under the Housing Grants, Construction and Regeneration Act 1996 (HGCRA));
- an unreasoned award.

2.3.1   *Awards on jurisdiction* (section 31(4))

The operative part of an award dealing solely with the matter of jurisdiction is a ruling that the arbitrator does (or does not) have jurisdiction in the identified issue(s). Where the matter of jurisdiction is contained within the award on the substantive dispute, it is appropriate that such finding is set out early in that award. That is likely to be either immediately following the recitals as a whole, or immediately following those recitals which relate to the objection to jurisdiction, before proceeding with any remaining recitals.

Some people have difficulty in distinguishing between 'jurisdiction' and 'power'. Jurisdiction establishes the limits within which power(s) can be exercised. This can be illustrated by reference to sports activities. There are two adjacent football pitches. Two referees are appointed, one to act in each match. Both have identical powers – those granted by the rules of the relevant sporting body. Each can exercise those powers *only* in respect of the match he has been appointed to supervise. Each has *jurisdiction* over one specific match and can exercise his *powers* over that match but obviously not over the other.

One or more of the parties to an arbitration agreement might raise objection as to the jurisdiction of an arbitrator. Likely situations where this can arise are:

(a) The arbitral appointment has been by an 'appointing authority'. After such purported appointment has been made a party contends that the dispute falls outside the terms of the arbitration agreement. That party might, for instance, contend that '...the arbitration agreement provides that all disputes arising *during* the provision of the services be referred to arbitration...', but this dispute first arose after the completion of those services.

The arbitration agreement does not therefore apply to this particular dispute.' They thereby contend that the appointed 'arbitrator' does not have jurisdiction.

(b)   Regardless of whether the appointment has been made by the parties or by an appointing authority, party A has made a claim against party B. It is in specific terms (X). They cannot agree and appoint an arbitrator (or apply for such appointment and one is made). The arbitrator orders exchange of pleadings or of statements of case. The claimant's statement includes a matter (Y) which, whilst now in dispute, was not in dispute at the time of the appointment – or, if it was, it was not mentioned in the document(s) requesting the appointment. The respondent therefore contends that whilst the arbitrator has jurisdiction in respect of dispute X he has no jurisdiction in respect of dispute Y.

Under sections 30/31 of the 1996 Act the arbitrator may rule on an objection to his jurisdiction in one of two ways:

(a)   in an award as to jurisdiction; *or*

(b)   in the award on the merits of the case (i.e. in the substantive award).

Note also the effect of section 32(4).

In both instances, by section 52(4), the award on jurisdiction should give reasons.

2.3.2   *Agreed awards* (section 51(2))

The operative part of an agreed award is, in essence, virtually identical with that which would have applied had the award been entirely the arbitrator's decision. The main distinction is in the recitals, where the fact of the settlement is set out.

Section 51 of the 1996 Act now makes provision, unless otherwise agreed by the parties, for such settlements to be recorded 'in the form of an agreed award', provided that the arbitrator has no reasonable objection. That award has to 'state that it is an award of the tribunal'. In whatever manner the arbitrator is informed of such settlement, he must ensure that the notification of that settlement, or the draft award embracing it, is signed by an authorised representative of *both* parties.

In some instances the parties' representatives, particularly if they are lawyers or have had little experience of that particular arbitrator and his ability, will agree a complete draft (leaving the arbitrator to do little more than to insert the amount of his fees, sign it and have his signature witnessed).

In other instances the representatives (or the parties themselves) will simply notify the arbitrator of the bare settlement details, leaving it for him to draft the award as a whole, incorporating those settlement terms, and then obtain the parties' agreement to that document.

In the event of an agreed award, the parties know their reasons, so (unless the parties agreed otherwise) reasons would not be given (see section 52 (4)).

2.3.3 *Awards on separate issues* (section 47)

The operative part of an award on a separate issue is, in essence, the same as if that had been the only issue and thus as described for an award of that type (monetary, performance, etc.)

There can within a single dispute be several, sometimes many, disparate issues – and sometimes there can be advantage in having separate awards for single, or compatible groups of, issues. This arrangement sometimes stems from adoption of the most convenient manner and sequence of dealing with submissions and hearing(s), sometimes from the possibility that an award on one or more issues will facilitate a settlement on others.

If the parties agree of their own accord that a series of awards on different issue are to be made, then (unless the arbitrator persuaded them that they should agree otherwise) that is the sequence the arbitrator must follow. If no agreement has been made by the parties on this point, it is for the arbitrator to exercise his discretion, having regard to his duty under section 33.

In this and other instances where there is to be more than one award, each such award should recite that intention. The recitals in the later awards should only repeat those which appeared in the first award as necessary to give the background to the current award, the remaining recitals being incorporated by reference. Each award must identify which issues are dealt with in that award. Other than the last in the series, each award should specifically reserve the remaining matters to a further or final award (usually as the last-placed item, before the signature). There can then be no doubt as to which awards are not, and which is, the final award.

2.3.4 *Awards on reserved matters* (equivalent to section 47)

The operative part of an award on reserved matters is in most instances similar to that in the appropriate remedy-based awards referred to earlier. For instance, a reserved award on liability for costs (i.e. excluding the amount of such costs at that stage), is similar to a declaratory award. One on the quantum of costs is similar to a monetary award.

This type of award arises only where its subject matter has in an earlier award specifically been reserved to a further or final award. It is not to be confused with a corrective or additional award under the provisions of section 57 of the 1996 Act (and see section 2.3.5 below).

Reserved matters might be further substantive issues, or might be, or might include, interest where party submissions have been necessary, following an award on the main monetary decision. The matters reserved might be costs – either dealing with liability for costs or, where that has already been done, dealing with the assessment of the amount of recoverable costs (what used to be referred to as 'taxation of costs').

The requirement for reasons applies equally to such an award. The fact of the matter having been reserved from an earlier award should be recited in the early part of the award. Otherwise, the position regarding recitals is as described under 'Awards on separate issues' at 2.3.3.

2.3.5   *Corrective or additional awards* (section 57)

The operative part of a corrective award sets out (or adequately identifies in the preceding text) the correction and specifies that it is to be read as incorporated in the earlier award. The operative part of an additional award is of a similar nature to what has been described for an award (be it monetary, declaratory, etc.) dealing with that type of issue. It must, additionally, specifically and unambiguously make the correction or addition.

Under the provisions of section 57, unless agreed otherwise by the parties, the arbitrator may (within the time limit laid down):

(a)   correct an award in which he has made a clerical slip or error of the kind described in the section, or clarify or remove any ambiguity in the award; *or*

(b)   make an additional award where a presented claim had not been dealt with by him.

Awards under this head are to be distinguished from such awards made as a result of the original award being remitted to the arbitrator by the court following application to the court. Those awards are considered under the next heading.

A correction might relate to a matter of recital or of analysis, or might relate to the original operative part as such. In both instances, if the underlying award was to give reasons, the correction or addition must likewise give reasons, but cannot be used to change the original intention. Both corrective and additional awards may be made on the arbitrator's own initiative or following the application of a party to that effect. Once faced with making a correction, the arbitrator should, as a matter of prudence at least, check the whole award for any further clerical mistakes or errors of the type covered by that section – and deal with them.

The matters giving rise to the corrective or additional award should be recited in the early part of the award. Otherwise, the position regarding recitals is as described under 'Awards on separate issues' at 2.3.3.

2.3.6   *Awards following remission* (see sections 68(3)(a) and 69(7)(c))

The operative part of an award following remission will be of a similar nature to an award of that type, i.e. monetary, declaratory, etc.

Under section 68 of the 1996 Act a party can challenge an award on the grounds of what is there described as 'serious irregularity' and under section 69 can appeal on a point of law. One of the avenues open to the court is to remit the award, or part of it, to the arbitrator for reconsideration. Under section 69, that reconsideration is to be 'in the light of the court's determination', i.e. giving proper regard to the decision of the court on the point or points of law referred to it.

On such remission, the arbitrator must reconsider the matter(s) remitted. If under section 68, he must correct it so as to remove the effect of the relevant and identified 'serious irregularity'. If under section 69, he must make such amendment as is generated by the court's determination on the point of law. In either event he must, by section 71(3), make a fresh award on the remitted matters and must do so within the time laid down.

The award should deal with resultant costs of the arbitration to the extent that (if such be the case) the court's order does not do so, or must adopt the court's direction as to costs.

Again, the matters giving rise to the remission should be recited early in the award. Otherwise, the position regarding recitals is as described under 'Awards on separate issues' at 2.3.3.

2.3.7   *Awards giving further reasons* (see sections 52(4) and 70(1), (4) and (5))

If the reasons given in response to an order of the court (or a request of the parties) are published as a separate document (i.e. not by way of a corrected inclusive award) the document should include a declaration by the arbitrator that such reasons relate to and form part of the previously-issued award, making it clear whether they are the entire reasons, or are additional reasons. If the reasons are given by way of a replacement award, now completed by the insertion of reasons, there would usually be no change to the operative section.

Such an award, in either form, can arise in two ways:

(a)   Where one or both parties do not (or profess to not) understand how an award was arrived at – because it gives no reasons, or the reasons are or appear to be inadequate, or do not support the conclusion. One or both of them might request the arbitrator to remedy the deficiency.

(b)   Where one or both parties apply to the court under section 68(2)(h) (failure to comply with the requirements as to the form of the award).

(c)   Where the court orders under section 70(4) that the arbitrator shall state his reasons in sufficient detail (for application or appeal).

If, in order to enable it to consider the application or appeal, the court orders that the arbitrator states reasons in sufficient detail for the court's purpose, the court may also make an order with respect to the additional costs incurred. The arbitrator must give effect to such order.

The document should recite how and why it came to be made. Otherwise, the position regarding recitals is as described under 'Awards on separate issues' at 2.3.3.

2.3.8   *Awards following payment during the reference* (all as 'substantive' monetary award, plus section 49(3)(b))

This is simply a variant on a 'substantive' award, included here for convenience of cross-reference to later illustration(s).

The operative part of the award would be similar to a monetary award, but would also take account of the payment made – by adjustment of the sum otherwise awarded. The method of expressing that could be dependent upon how pleadings had been amended in consequence of the payment.

The fact of the payment would be recited, along with the date and any applicable conditions. Section 49(3)(b) makes provision for the award of interest on such payment, or any part of it.

2.3.9   *Awards following:*

2.3.9.1   an earlier order giving provisional relief under section 39, *or*

2.3.9.2   an earlier adjudicator's decision (e.g. under the Housing Grants, Construction and Regeneration Act 1996)

For present purposes both these are considered only in circumstances where the preceding order or decision had ordered the payment of money. If a relief or remedy other than monetary had been ordered in either instance, the subsequent submissions and the award would need to 'unscramble' the previous decision if it differed. That would extend into areas beyond the scope of this book.

The operative part of the award would again be as the relevant 'substantive' award, but would make clear how the amount paid had been dealt with. The manner of expressing that would depend upon whether the amount claimed in the arbitration was a gross figure or was a net figure after deduction of the amount already paid.

The fact of the provisional order or adjudicator's decision would be recited, as would the payment. Generally, the award would follow a pattern similar to 2.3.8 above.

2.3.10   *Unreasoned awards* (section 52(4) exception, if the parties so agree)

The operative part of an award without reasons is effectively the same as in one with reasons.

Under the provisions of section 52(1) 'The parties are free to agree on the form of an award' and by section 52(4) the parties can agree to dispense with reasons. An award without reasons still needs to recite at least sufficient to identify the parties, how the arbitrator came to have jurisdiction, what was in dispute (for the parties might have several current disputes of which this is one) and the fact that the parties agreed to an award without reasons.

2.4   'Procedural' awards

As defined for the purposes of this book, these are:

- an award dismissing the claim;
- an award on abandonment.

2.4.1   *Awards dismissing the claim* (sections 41(3) and (6))

The operative part of an award dismissing a claim is a determination and declaration making it clear that the claim is dismissed. If there are several issues and it is only in respect of one issue (or others which are identified) that the dismissal decision is made, that claim (or claims) is identified and dismissed, the other issues remaining intact and being dealt with in the normal manner.

Unless the parties have agreed otherwise, an arbitrator can make an award dismissing a claim, under the provisions of section 41(3) ('inordinate and inexcusable delay' (as there qualified)) and section 41(6) ('claimant fails to comply with a peremptory order ... to provide security for costs' – by section 82(1) 'claimant' includes 'counter-claimant').

The matters giving rise to the dismissal should be recited, thus in effect giving the reasons.

2.4.2 *Awards on abandonment*

The operative part of an award on abandonment (provided that an award is appropriate, see below) is in effect a declaratory award which makes it clear that the parties have abandoned the reference – and declaring the arbitrator *functus officio*. The terms of the declaration would depend upon the background to the abandonment (see below).

The parties might simply cease to take any action in connection with the arbitration (in which case the arbitrator should seek to ascertain their intentions), or both might inform the arbitrator that the arbitration has been abandoned, or that neither side will be taking any further steps, or something of that kind. If it appears possible that they have reached some kind of compromise, the arbitrator should usually check whether they wish him to issue an agreed award putting such settlement into effect (see 'agreed awards' in section 2.3.2 of this chapter).

They might say that they are content to let matters rest as they are, or that all matters under the settlement have been concluded and nothing remains to be done, by them or by him. In that event he can inform them that if he does not hear from them within a reasonable period, which he fixes, notifying him of details of the settlement, he proposes to issue an award. He would indicate that such award would be to the effect that the matters in dispute in the arbitration had been settled by mutual consent in terms which the parties have withheld from him – and declaring himself *functus officio*. Such award would deal with his own fees. Should he still not be informed of the settlement terms he could issue such an award, thus closing the matter. In it he would also briefly recite why and how it had become necessary to close the matter in that way.

The parties might not reply at all. The arbitrator needs to ascertain the true position if he can do so by reasonably practical enquiry of them. His enquiry should inform them of what he proposes to do in the event that they both confirm that the position is as in the 'content to let matters rest' situation. It should also inform them of his proposed action if he receives no reply. If he receives that confirmation he can issue the kind of award described above. If he receives no reply he can seek to give finality by issuing a 'closing order' (see 8.2.2.2(1)), in very similar terms to such an award. There is no statutory provision for such a document, but its purpose is clear and such a device has long been adopted by some arbitrators.

Whichever course is followed, the manner in which it arose should be recited.

2.5 'Institutional' awards

As defined for the purposes of this book, these are:

- an award under rules or under other statutes;
- an award under 'consumer schemes'.

2.5.1 *Awards under rules* (section 4(3)) *or under other statutes* (section 107 and Schedules 3 and 4)

The purpose of mentioning such rules and statutes here is simply as a reminder that an arbitrator must ensure that he complies with any requirements relating to awards which might be included in such rules (where adopted) or applicable statutes.

The party autonomy provisions (their freedom to agree various powers and procedures) can be put into effect by adopting institutional rules published by a variety of trade and other bodies. Such rules can affect the form of an award to the extent that they incorporate what amounts to party autonomy under section 52 of the Act. They can also affect the recital of, and decisions upon, such matters as procedures, interest, costs and so on.

Similarly, various statutes dealing with a wide variety of matters (e.g. the Agricultural Holdings Act 1986) include provision for certain disputes to be resolved by arbitration. It is possible that such a statute might limit the nature of dispute which can be so referred, or impose limitations or other requirements which can influence the content and format of the award. Where that applies, those provisions must be followed.

2.5.2 *Awards under 'consumer' schemes*

The requirements of an enforceable award described previously apply equally to such awards, subject to the provisions of the rules of the scheme. An arbitrator acting under these schemes should be aware of sections 89–91 of the 1996 Act (Consumer arbitration agreements).

There exist a number of arbitration schemes in what is generally termed the consumer field – schemes relating to holidays, to personal insurance, to the installation of glazing; a wide range of provider/personal customer situations. They are all subject to specific rules which form part of each scheme. Those rules can include provisions which govern various aspects of the resultant awards, not least the costs of the arbitration and how they are to be dealt with.

2.6 'Ancillary' awards

2.6.1 *When is an award (perhaps) not an award?*

There are situations where what is sometimes loosely referred to as an award is not, or is not always, or is not of necessity an award. Cases in point are decisions under:

(a)   section 39 (power to order relief on a provisional basis), which applies *only if* the power is conferred by the parties; *and*

(b)   section 38(3) (power to order the provision of security for costs), which applies unless the parties have agreed otherwise.

Although section 39 has a side heading referring to provisional *awards*, the view of many is that the title is an historical accident stemming from positional changes as the Arbitration Bill proceeded through its various stages, and that the provision is for an order. That interpretation is preferred here, but some commentators take a different view.

Section 38(3) has no such reference to an award, but nonetheless there are some who contend that they would wish to have, or to publish, such a decision as an award.

To avoid blurring between procedural orders and awards or expanding into realms beyond the parameters of this book, these two provisions are not developed here, although it is recognised that party agreement could provide for the decision(s) to be by way of an award – with all the ramifications of enforcement and appeal.

However, if an award as such is required by the parties and the arbitrator complies, the essentials of an enforceable award apply equally to such awards. In addition, see the note at the end of 8.4.

3   Style, content and checklists

3.1 Style

3.1.1 Narrative style

Proided that the resultant award complies with the requirements which have been considered earlier, there is no mandatory style for an award. Each arbitrator should adopt a style, format, sequence and form of language with which he is comfortable and which is most appropriate to:

(a)   that particular arbitration; *and*
(b)   the particular parties.

There is, however, a well-recognised general approach – at least a basis of sequence and nature of content, from which individual styles or particular awards can evolve. Some arbitrators, with the confidence of experience behind them, adopt a freer and more flowing style, some a minimalist or staccato style, some adopt a kind of legalese; some will prefer a different sequence.

In determining the narrative style to be adopted, an arbitrator must have in mind the potential readers – principally the parties and their advisers. The award might also be read by a judge should the need for enforcement or appeal arise. The story which the award briefly tells and then resolves must be clear not just to the arbitrator who wrote the award, but also to other potential readers. That clarity must extend to all aspects of the award, including (unless the parties want an award without reasons) showing how the decision was reached.

Several decades ago it was commonplace, almost universal, for arbitrators to preface each item in the non-contentious introduction with the word 'Whereas', later restricted to the single use of the word at the beginning of those items, thus relating to them collectively. Some arbitrators then dropped that and replaced it with the heading

'Recitals'. Many then abandoned that in favour of 'Background' or 'Introduction', or other suitable heading.

The use, or not, of 'whereas' is a matter of individual taste, or of established use in a particular area of activity. The same applies to the use, or not, of a heading which embraces the whole of the non-contentious introduction, be it 'Recitals', or 'Background', or simply the use of sub-headings relating to each of the sub-topics, in whatever sequence best suits the case. Such a sub-heading might be, say, 'The Parties' or 'The Parties and their relationship', and so on.

It is imperative that the content of that introductory section of the award is factually correct and does not meander into other matters such as analysis, or findings of fact, relating to the matters in dispute.

The style and format outlined in the following chapters should provide a useful foundation from which newcomers to the task can develop a personal style, or from which to create a personal checklist or series of such lists, and from which to develop their preferred sequence, whether in general or for a particular award.

3.1.2 Language

'Language' here refers to vocabulary and mode of expression. Additionally, the submission might require that the award be in one or more national languages (section 34(2)(b)).

The vocabulary and mode of expression should have adequate regard to the nature of the dispute and of the parties. It should be readily comprehensible to the particular parties and be unambiguous. To the extent possible it should be in plain English. An award in a commercial dispute will usually be expressed in normal commercial English, the kind of narrative used in a well-prepared commercial report. An award in a dispute concerning some complex scientific or other specialist topic is likely to use at least some specialist vocabulary. An award involving a minor quarrel between domestic neighbours should, to the extent possible, be couched in language which is understandable to those parties.

Where there is little or no submission on points of law, the use of legal terms might well not arise. Where the dispute is substantially about legal interpretation, the submissions will be heavy with legal terms and the arbitrator will make a decision based upon his analysis of those legal submissions. In consequence he will in his award adopt the appropriate legal terms from those submissions – whether he is a lawyer or not. That can be a pitfall for the unwary or the rash arbitrator. He might reach a valid decision but if he uses mistaken legal language, it might open up the award to challenge. In particular, a non-legally-knowledgeable arbitrator should avoid importing, *of his own volition*, legal terms which have not been used before him by the parties or their representatives. By way of example, for a non-lawyer it is preferable to say something like 'The Respondent acted in a manner whereby he led the Claimant to believe that the Respondent would not expect the Claimant to comply with the formalities of giving written notice of...' rather than to use terms pertaining to, say, 'waiver' or 'estoppel' where such terms had not been used by the parties.

3.1.3   Headings and list of contents

The use of headings which split the award into functional parts can be of considerable assistance to a reader. They can also be a useful device for the drafter, not least for use as a specific checklist or for comparison with a prior-produced list. (See 3.2 and 3.3 below.) The nature of the headings is likely to vary from case to case.

If an award amounts to only a few pages, a list of contents could be entirely superfluous. In long or complex awards, however, a bound-in list of contents is of potential assistance, for the arbitrator in maintaining the logic and the narrative flow of the award, and to help the parties or others reading the document, both in following that logic and in finding sections when later referring back.

3.1.4   Reasons and their incorporation

Reasons for an arbitrator's determination are his decisions as to disputed facts, his consideration of the law or applicable basis[1] for determination, and his application of the latter to the former.

Reasons can be included in a variety of ways: (a) by being set out in a section of the award entitled 'Reasons', either within the body of the award or as a cross-referenced annexe, or (b) by incorporation in the arbitrator's analysis, following in sequence after the recitals and thus leading a reader logically to the arbitrator's decision. That sequence is adopted in the illustrations later in this book, other than a brief illustration at 9.3.7.2.

Some methods of showing reasons follow the established tradition (if there is one) in the relevant area of arbitral activity. Some follow the requirements set down in any applicable rules or by agreement of the parties. Some stem from the arbitrator's preference.

3.1.5   Checking

Finally on this topic, the arbitrator should always re-read the award before finalising it and publishing it, and similarly with any early-drafted sections. He should check that every issue to be dealt with in

---

[1]   Section 46(1) of the 1996 Act permits the parties to agree that either (a) a stated law or (b) 'other considerations' is to form the basis of the arbitrator's determination.

that award has been dealt with, that the incorporated reasons are clear, and that any factor which is necessary to give credibility and lucidity to the award is recited, as succinctly as possible. He should attempt to put himself in the position of a stranger to the case and consider whether such a person would properly understand what the award was about and, if reasoned, why it came to that decision. Such a precautionary step does not only apply to the decision itself, but also to the matter of style – of vocabulary, of syntax. It also applies to ensuring that those 'recitals' do adequately identify whatever should be identified. (See 3.2 below.)

3.2 Content

3.2.1 *In general*

An award should, usually first, inform a reader of:

(a) how and why the arbitrator comes to be resolving the dispute and with what authority;
(b) any restrictions on or expansions of his powers;
(c) any procedures he is required to follow, or has opted to follow under his default or party-agreed powers.

Those matters should logically be placed at or close to the beginning of the award. That inclusion tends to make an award longer than would be a judgment dealing with the same issues.

The award can then proceed in a manner which might be not unlike a judgment, setting out:

(a) what is in dispute;
(b) what was contended by the parties.

That in turn can be followed by:

(c) a brief precis of the relevant evidence adduced and submissions made;
(d) the arbitrator's consideration of the submissions and evidence; and (sometimes separately, sometimes merged) his conclusions, i.e.:
   • what he has found as (i.e. has determined to be) facts;
   • what he has held as matters of law (or, if section 46(1)(b) applies, 'other considerations');
   • his application of that law (or 'other considerations' to the facts he has found.

That leads to an unequivocal statement of:

(e) what has to be done (or foregone) by whom, or what his declaration is.

Whilst the 1996 Act makes provision for various aspects of content, there is no standard sequence nor phraseology for a valid award other than any which might be demanded by:

(a) any agreement under section 52 of the 1996 Act;
(b) the terms of reference;
(c) any relevant rules, or the like.

3.2.2 *Basic structure*

Although the structure and narrative form adopted by many experienced arbitrators vary in sequence and in detail, there is fairly wide consensus on a broad general basic structure for an award – at least to the extent that such structure gives a foundation for expansion or adaptation. Such basic or underlying structure is illustrated here. There will in some instances be a different grouping or sequence better suited to the particular arbitrator's thought process, or to the particular case, or both. Arbitrations can arise in a variety of ways, some of which can affect the sequence of an award, all of which will affect content. For the purposes of this book, a basic checklist is explored, then fleshed out in the illustrations in later chapters.

The following list illustrates a possible broad basic structure such as that mentioned above:

A   Heading;
B   Background:
   B1   Identification and jurisdiction;
   B2   Interlocutory procedural matters[2];
   B3   Hearing;
C   Submissions and evidence/result of any inspection(s);
D   Analysis/findings/reasons/decisions, *issue by issue*;
E   Value Added Tax implications;
F   Interest;
G   Costs;
H   'Operative' section (i.e. that giving effect to the award);
I   Reserved matters;
J   Signature and formalities.

The splitting of this list into sections A–J for the purpose of this book is not intended to imply that an award should be similarly split or referenced. It is done here and in the illustrations simply to facilitate later cross-referencing.

[2]   'Interlocutory procedural matters' relates to matters arising during the course of the arbitration but incidental to its main purpose, being the publication of an award.

The above checklist is expanded and elaborated below at section 3.3, but the reader is reminded of the caveat that different circumstances can warrant different content and sequence. In some situations, for instance, certain matters can better explain the decision which follows if they are recited in the 'Analysis' section (D) rather than in either the 'Background' section (B), or the 'Submissions and evidence' section (C). That sequence can be useful where certain procedures or powers apply to one or more issues only and not to the others. Similarly, sections B1, B2 and B3 are readily identifiable parts of the background (B) but, depending upon circumstances, might be wholly or partially incorporated elsewhere if more convenient.

Each arbitrator should produce his own checklists, one as a basic overall reminder for general use, then a condensed (or expanded) one tailored to each particular arbitration or type of arbitration. The latter can be used to generate the appropriate sequence and content to be the objective in the particular case, then as a cross-check on completion of the award. In a case with very few procedural complications the simple list shown above might suffice; for more complex situations a more comprehensive list is necessary. The list shown and explained below is just one possibility.

3.3   A possible basic overall sequence and expanded checklist

This expanded checklist forms the core, or skeleton, upon which succeeding chapters are based. It is further explained in Chapter 5 by narrative interspersed with illustrations, which collectively form a monetary award. For that purpose, and for the purpose of linking illustrations in subsequent chapters back to that explanation, an alpha-numeric referencing system has been adopted for the checklist. These references are used as a supplement to, but separate from, the decimal enumeration used for sections of the book. As an example, a cross-reference to 'orders re language and translations' (see the checklist at 3.3.4) would be B2(i)6.

For ease of subsequent reference, the illustration given here is repeated without the intervening narrative as Appendix 1.

3.3.1   Section A Heading

Logically, the award starts with a suitable heading, which commonly includes (but see chapters 5–9 for phraseology):

A
(a)   the applicable Arbitration Act;
(b)   any other relevant statute, or scheme of arbitration, or rules;
(c)   identification of the parties.

Some arbitrators follow that with a short preamble giving a brief historical outline of the events which led to the arbitration. Others prefer that outline to develop out of the information recited under the following or similar items. The latter course is followed here.

3.3.2 Section B Background

(or 'Introduction' – or, if still preferred, 'Recitals' or other heading.)

As described earlier, there can be sub-headings for the various topics addressed, instead of, or in addition to, the heading 'Background'.

Note particularly that the number of items shown under this heading does *not* imply that this section of the award should be lengthy or form a major part of the award. It should be as short as reasonably possible. Not all of these items will apply in any one arbitration or any one award; they are here solely as the bones of an extended checklist. Recital of contentions, procedures, applications and so on should be succinct and be restricted to those which have a direct bearing upon the award. Evidence as such would not usually be recited at this point.

Some checklist items are intentionally repeated in later sub-sections under this 'Background' heading. That repetition is done primarily as a function of a checklist as such, rather than as an indication of sequence; for powers of the arbitrator, or procedures, can be established or changed by the parties at various junctures, earlier or later in the reference. Specific mention of those powers or procedures, whenever established, should be made only to the extent that they are relevant to the particular award. Some items can be transposed to later sections if that benefits the natural flow and overall clarity of the award.

At whatever stage unusual or relatively unusual powers might have been granted to the arbitrator, and in whichever manner he records them in the award, he should also record whether he has used any of those unusual powers (or procedures), in respect of what matters he has used them, and in what way.

The following are the background items referred to above, split into three logical sections. They will not always strictly follow this subdivision or sequence and this list is not necessarily comprehensive.

3.3.3 Section B1 Identification and jurisdiction

This section can conveniently summarise the matters which led to, and at the outset govern, the appointment of an arbitrator for the particular reference.

For each individual arbitration, the arbitrator should decide which, if any, of the following items need to be recited, which do not apply and can be deleted, and which can to better effect be included in a later section. He should then create a specific checklist and transpose those identified items into the appropriate later section.

B1

(a) identification of the parties;

(b) identification of the contract, or other relationship between the parties;

(c) the law governing the contract or other relationship;

(d) the essential provisions of the arbitration agreement/arbitration clause;

(e) the provisions for appointment of an (the) arbitrator(s) (including any special qualifications of, or requirements relating to, appointees – and any incorporated agreement on powers and procedures);

(f) any rules incorporated from the outset (section 4(3));

(g) the fact of a dispute having arisen; any known counterclaim;

(h) the *general* nature of the dispute(s) and the matters in issue;

(i) the resultant appointment – and any necessary details and terms;

(j) any challenge to jurisdiction and how dealt with (sections 30–32);

(k) basis for deciding the dispute (section 46);

(l) other matters to the extent they are relevant to this sub-section.

3.3.4 Section B2 Interlocutory procedural matters

This section can conveniently summarise the relevant procedures which arose or applied during the subsequent stages up to the hearing. Their inclusion should be restricted to the extent that these matters impact on or are necessary to explain or to give authority to the award.

Again, for each individual arbitration, the arbitrator should decide which, if any, of the items which follow need to be recited, which do not apply and can be deleted, and which can to better effect be included in a later section. He should then create a specific checklist and transpose those identified items into the appropriate later section.

Whether in this section or transferred elsewhere, the best sequence will often vary from that shown here. For example, relevant action or orders, whenever arising, are sometimes best recited as one sub-section of the award. In other circumstances they are best, or more conveniently, recited when referring to the particular meeting or stage of the proceedings to which they refer.

For brief explanation of the term 'costs in any event', see Section G at 3.3.10.

B2

(a) reminder to transfer appropriate items (if any) to a later section;

(b) the 'seat' of the arbitration (sections 3 and 52);

(c) any interlocutory agreement(s) regarding adoption of rules, or on powers or procedures (sections 4(3), 34–41), e.g., (but not restricted to):

1 on procedural and evidential matters (section 34) (insert detailed sub-list if appropriate);

2 on consolidation or concurrent hearings (section 35);

3 on appointment of experts, legal advisers or assessors (section 37);

4 relating to property the subject of the proceedings (section 38(4));

5 on preservation of evidence (section 38(6));

(d) if appropriate, which 'uncommon' powers have been used, and how;

(e) any necessary clarification or amendment of the matters in issue;

(f) any previous award(s) and what issues were decided by each (section 47);

(g) the issue(s) (claim and any counterclaim) to be decided by this award;

(h) meetings: preliminary, further, pre-hearing review;

(i) consequent and subsequent applications and orders, re (e.g):

1 attended hearing, or 'documents only' (section 34(2)(h));

2 type of statements to be exchanged (section 34(2)(c)); timetable;

3 disclosure and production of documents (section 34(2)(d));

4 exchange of witness proofs and experts' reports (section 34(2)(f));

5 meeting(s) of experts (section 34(2)(f));

6 if appropriate, language and translations (section 34(2)(b));

7 questions to be put, when and in what form (section 34(2)(e));

8 other relevant matters;

(j) resultant party action:

1 brief details of exchange (section 34(2)(c));

2 admissions or agreements; result of meeting(s) of experts;

3 any resultant reduction in, or expansion of, the matters in issue;

(k) other directions and/or administrative actions if, and to the extent, not already outlined and if relevant to the award, for example:

1 orders for security (section 38(3) and 41(6));

2 limitation of costs (section 65);

3 any valid 'costs in any event' agreement (section 60);

4 any 'costs in any event' directions (section 61(2) exception);

(l) any provisional orders issued under section 39 (if empowered);

(m) any party default (and here or later, the consequence), e.g:
  1 inordinate or inexcusable delay (section 41(3));
  2 failure to comply with any order or direction (section 41(5));
(n) issue of peremptory order(s), compliance or default (section 41(5) and (6)), consequence (section 41(6) and (7));
  (if appropriate, e.g. dismissing claim, repeated in operative section H); court enforcement (section 42);
(o) other matters to the extent that they are relevant to this sub-section.

3.3.5 Section B3 Hearing

This section can conveniently summarise the relevant procedures directly relating to the hearing, to the extent that they impact on or are necessary to explain or to give authority to the award.

Again, for each individual arbitration, the arbitrator should decide which, if any, of the items which follow can to better effect be included in a later section. He should then create a specific checklist and transpose those identified items into the appropriate later section. For instance, this sub-section can conveniently form an introductory part of Section C (Submissions and evidence) instead of being included here.

B3
(a) reminder to transfer appropriate items (if any) to a later section;
(b) date and place of the hearing (or if dealt with on the basis of documents and written submissions only, or on some other basis – the arrangements);
(c) failure to attend (or equivalent)/any ex-parte proceedings (section 41(4));
(d) who represented the parties (section 36); witness names and designations;
(e) whether oral evidence was taken on oath or affirmation (section 38(5));
(f) if not recited in B2, any default, and any orders issued, during the hearing;
(g) inspection(s) etc. (and who present) (section 38(4));
(h) any agreed (and final) list of issues; any payments during the reference;
(i) any other matters to the extent that they are relevant to this sub-section.

3.3.6 Section C Submissions and evidence

The background and general nature of the dispute having been adequately described, it can be convenient to set out a brief summary of the relevant parts of the parties' contentions, followed by the legal submissions and precis of evidence. This can provide a useful lead-in to the analysis which follows. The content and sequence will depend partly upon whether there is an attended hearing, or all submissions and evidence are committed to writing, or a mixture of the two. Evidence should only be recited where and to the extent that it is necessary for the proper understanding of the reasons which follow, whether in this or any other section of the award. It should never be repeated verbatim other than any essential, and usually very brief, extracts.

Where the arbitration is being conducted on a 'documents only' basis, or where complexity or other factors make it a suitable sequence to follow, it can be convenient to deal with these matters prior to proceeding to the analysis etc. which follows. Otherwise, and as is often the case, this section can be amalgamated with Section D, issue by issue.

A basic type of sequence might be:

C
(a) the contentions of the parties (claim and any counterclaim);
(b) a brief but adequate précis or summary of the respective advocates' opening and closing submissions, whether in writing, oral, or a combination;
(c) if not, or to the extent not, covered in Section B:
    1 identification of each witness, with status, qualifications, etc;
    2 brief precis of *relevant* evidence
      (including any admissions under cross-examination);
    3 equivalent description in ex-parte proceedings;
(d) if not, or to the extent not, covered in Section B, reference to whether the arbitrator had made any inspection of the subject matter or the place concerned ('the locus'):
    1 the purpose of the inspection and who present;
    2 if other than simply to put the matter into context, the result of inspection;
(e) any other matters to the extent that they are relevant.

3.3.7 *Section D Analysis/findings/reasons/decisions on the substantive issues*

It is important that the apparent brevity of this sub-list D is not misunderstood by a newcomer to the subject. In a case of any substance, by the time each and every issue of claim and counterclaim has been considered individually in respect of fact and law, however succinctly, this section will take up a substantial proportion of the award as a whole.

If in any specific area of trade or commerce there is an established method of dealing with reasons and their location in the award, that method would usually be adopted by an arbitrator acting in that field, provided that he was content that such procedure was valid and did not conflict with any party agreement. The same can be said for the overall format of the award.

A basic type of sequence for this part (whether separately or amalgamated with Section C) might be, with (b) and (c) repeated issue by issue of claim and counterclaim:

D
(a)   a numbered list of the issues dealt with in this award (unless sufficiently detailed in the recitals);
(b)   any common ground or undisputed facts;
(c)   an adequate, but brief, precis of and analysis of the submissions and evidence (unless it is to be an award without reasons), effectively thereby determining (in respect of liability and, if then relevant, the financial or other consequences) and giving the reasons for:
   1   findings of fact;
   2   what held as matters of law;
   3   application of the law to the facts;
   4   consideration of any interest claimed as special damages;
   5   the decision on the substantive issue(s),
       also showing why any contentions were rejected;
   6   consideration of whether a counterclaim is a set-off[3]
       (unless considered under Section G (Costs);

concluding with (leaving matters of VAT and interest, if applicable, and costs to be considered at Sections E to G below):

(d)   a summary of those decisions (if there are several to draw together);
(e)   where money is awarded, effect of any amounts claimed but paid during period of arbitration.

3.3.8 Section E Value added tax implications

An arbitrator has no power to determine whether, as between a 'taxable person' and Customs and Excise, VAT is payable. Matters relating to VAT can, however, arise in a variety of ways in connection with an

---

[3]   For present purposes a set-off can be considered to be a claim by a party which is in reality a defence to the claim of the other party, in whole or in part. Whilst a counterclaim can also be pleaded as set-off, a set-off pure and simple has no independent existence and cannot constitute a counterclaim.

award. So as not to generate inappropriate complications at this stage of the book, the matter of VAT, relevant definitions, and its various implications in respect of awards is addressed in moderate detail later at 4.3.2. For the present, a general reminder only is included in this checklist:

E
(a)   check whether there are VAT implications and, if not dealt with in 'pleadings', request parties' submissions;
(b)   seek party agreement on *manner* of treatment, including whether to be in present award, or to be reserved for later submissions;
(c)   adopt appropriate procedure (see Chapter 4, Section E).

3.3.9   Section F Interest

A basic type of sequence might be:

F
(a)   consideration (on claim and counterclaim) of any allowable interest to be added, the rate applicable and whether it is to be simple or compound:
      1   on any amount awarded (section 49(3)(a), or contractual, or at common law);
      2   on any amount claimed in, outstanding at the commencement of, and paid during the period of, the arbitration (section 49(3)(b)) – and see D(e) above;
      3   on the outstanding amount of any award (i.e. post-award interest – section 49(4)); (from the date of the award, or later date fixed, until payment – on outstanding amount of the award, on any interest awarded under section 49(3), and on any award as to costs);
(b)   decision(s), computations of amounts, and summary if appropriate.

That sequence is fleshed out in the list of relevant 'tasks' summarised at 4.3.3.2 and its preceding narrative.

3.3.10   Section G Costs

To a newcomer, the term used in section 61(2) of the Act 'costs follow the event' might be puzzling. The 'event' is the success (or not) of a claim. Whilst an over-simplification, it can be broadly translated as 'the loser pays the costs'. The matter is explored at 4.3.4.1.

'Costs in any event' refers to a situation where, regardless of the degree of overall success, the arbitrator has previously ordered (or the parties have validly agreed) that a party shall bear certain specified costs.

A basic type of checklist sequence might be:

G

(a)   consideration of liability for recoverable costs (arbitrator's fees and expenses, fees and charges of any arbitral institution concerned, legal or other costs of the parties) (sections 59–64);

(b)   consideration of validity of any agreement to pay costs in any event (section 60);

(c)   consideration of whether more than one 'event' (section 61(2);

(d)   if not dealt with earlier, consideration of whether a counterclaim constitutes a set-off (affects 'event': section 61(2));

(e)   principle adopted and reasons if not following the event (section 61(2));

(f)   basis of determination of recoverable costs (section 63(5));

(g)   effect of any 'costs in any event' orders or agreement;

(h)   effect of any limit on recoverable costs (section 65);

(i)   effect, and disposition, of any security provided (section 38(3));

(j)   effect of any earlier provisional order(s) for payment of costs (section 39);

(k)   effect of any rejection of offer in settlement of substantive dispute;

(l)   decision(s) on liability for costs – claim and counterclaim (section 61);

(m)   determination (if appropriate at this stage) of *amount* of recoverable costs (section 63(3), and, if not determined in principle at F(a)3 above, interest on costs (section 49(4)).

That sequence, too, is fleshed out in the list of relevant 'tasks' summarised at 4.3.4.2 and its preceding narrative.

3.3.11   *Section H Operative section (commonly headed 'Award')*

Whilst not essential, it is not uncommon here, as an introduction to this section, to mention the fact of having considered the submissions and so on:

H

(a)   the fact of having considered submissions;

It is then necessary to give precise conclusions and instructions to the parties, stating who has to do what and when, e.g:

(b)   1   if a monetary award, amount (sum plus any interest) to be paid (section 48(4));
        2   if a non-monetary award, the appropriate declaration, direction or, for example, rectification (sections 48(3) and (5));

(c)   any conditions or terms;

(d)   time for performance;

(e)   provisions relating to post-award interest (section 49(4)) – and see F(a)3 and G(m) above;

(f)   VAT implications if appropriate;

(g)   who is to pay what costs (section 61) (and, if a party has already paid any fees of the arbitrator(s) for which it is not liable, provision for reimbursement together with any interest relating to such reimbursement).

3.3.12   *Section I Reserved matters*

This requires a simple reminder:

I

(a)   Matters left over to a subsequent award.

3.3.13   *Section J Signature and formalities*

A list might be:

J

(a)   Date when award was made (and, if different, date when signed); seat, if not previously designated (sections 52(5), 53 and 54);

(b)   Signature (section 52(3));

(c)   Witnessed.

See 1.3.1.5 for effect of date of award. The matter of where made is addressed at 5.3.13.

3.3.14   *Final check*

On completion of drafting, the arbitrator should check that the requirements of an enforceable award set out at 1.3 have been met. It is highly desirable that the completed draft is put aside, preferably, depending on the degree of urgency, until at least the next day, but avoiding undue delay, then perused again with a fresh mind and eye.

4 An approach to decision-making

4.1 Background

Comments and illustrations relate to English law and are on the basis that, unless stated otherwise, all 'default' provisions of the 1996 Act apply without alteration and that there are only two parties.

The previous chapter has in general terms considered a possible format and sequence of an award – the *how?* and *where?* Before moving to illustrations of awards, one must consider means of arriving at the *what?* – i.e. the decisions which are to be set out in the award. This chapter looks at possible procedures or tasks involved in that decision-making process, i.e. the *management* of decision-making relating to the substantive issues and to matters which stem directly from those decisions. If the decision-making process is well-planned and logical, it should lead to a well-planned and logical award.

Any substantial excursion into the law of contract, of tort, of evidence, of the law relating to interest and to costs, and other legal matters, would stray from the purpose of this book. Those topics should be studied separately. However, in order to put some of the decision-making tasks into context and to give them substance, part of this chapter briefly expands upon those subjects.

As a starting point for that part of the decision-making process addressed here (i.e. that which principally relates to the substantive issues and VAT, interest and costs), it has to be assumed that except where the context indicates otherwise:

(a) the parties' submissions are clear, unambiguous, well cross-referenced, and whether by common consent or by order of the arbitrator, precisely identify the issues and the contentions of the parties;

(b) the arbitrator has so indexed his evidence books as to enable him rapidly to turn up his notes on any issue or sub-issue, i.e. his notes recording the oral evidence of each witness (indicating whether this was in examination-in-chief, or cross-examination, or re-examination), any oral submissions by the advocates and any relevant interventions or applications;

(c) the arbitrator, whether manually or by the use of information technology has, issue by issue – and in a complex case, sub-issue by sub-issue – collated the location of oral and documentary evidence and submissions, including opening and closing addresses (see under 4.3.1.2 and Appendix 2).

If that has been done, the arbitrator is in a position readily to re-absorb and to compare all relevant matters submitted to him on each individual topic.

4.2 Underlying matters

The underlying requirement is that the decision, and the subsidiary decisions leading to it, must be the arbitrator's own. He cannot delegate that judicial function. If, for instance under section 37 of the 1996 Act, he has appointed an assessor to assist him on technical matters he must:

(a) give the parties a reasonable opportunity to comment 'on any information, opinion or advice offered by...such person'; *and*

(b) having considered what the assessor has had to say and the parties' comments, make up his own mind. In doing so he must avoid giving himself 'own evidence' (i.e. he must not use his own knowledge or expertise without notifying the parties and giving them an opportunity to address him on the points identified).

Matters on which the arbitrator must be clear and, to the extent that they are relevant, should list for his own guidance before he starts the decision-making process include, but are not restricted to:

(1) Is the arbitration subject to any institutional or other rules? If so, do such rules impact upon the nature of the award and/or on the decision-making process? Do they impose limits of any kind? Or restrict the nature of submissions? Do they expand or restrict the powers of the arbitrator? Do they make provision relating to costs of the arbitration? And so on...

(2) Have the parties by agreement expanded or restricted or removed any of the default powers of the arbitrator under the 1996 Act?

(3) Has there been any challenge to jurisdiction? If so, has this been resolved, or is it to be resolved in this award?

(4)   Have the parties under the autonomy provisions of that Act made any agreement as to procedures? Or as to the nature and interpretation of submissions (for instance, is silence in a response to a statement of case to be deemed to be acceptance of a contention)? And so on...

(5)   Have the issues been pre-defined? (For instance in the wording of a 'submission agreement', i.e. an agreement referring an existing dispute to arbitration, or in a joint invitation to act, or in an application for an institutional appointment.)

(6)   Are the strict rules of evidence to apply (at this stage regarding the use the arbitrator makes of and weight given to the evidence)? If not, have any criteria been established (other than the section 33 requirement of fairness and impartiality)?

(7)   Is the determination to be based upon other than a stated law? If so, is the basis that of equity, trade practice, or something else? If something else is that within the bounds of section 46(1)(b) and the 'public interest' provisions of section 1(b)?

(8)   Have the parties pre-agreed the form of the award and/or whether there is to be more than one award?

(9)   Have the parties in any way expanded or restricted the arbitrator's default powers:
      (a)   regarding remedies, under section 48?
      (b)   relating to the award of interest, under section 49?

(10)   Have the parties made any valid agreement regarding liability for costs of the arbitration under section 60, or as to the principle to be applied in determining liability under section 61(2), or as to what are to be recoverable costs under section 63?

(11)   Has any limit on the amount of recoverable costs been imposed under section 65?

(12)   Have any 'costs in any event' orders been made?

(13)   Are there any other factors which do, or might, affect the nature or preparation of the award?

(14)   Are all the original matters referred to arbitration in this reference still in issue; have any further issues been incorporated into the arbitration by party agreement; have any been withdrawn or settled?

There can be other factors too, both in general and within any particular area of commerce or other relationships.

The arbitrator must give full effect to all of these and any other underlying matters when addressing the issues before him. They arise again in the 'tasks' listed under Sections D to G of an award later in this chapter.

For each award, in preparing to consider the issue(s) and reach his decision on them (combining those matters upon which the parties have agreed and those left to his discretion), the arbitrator must:

(a) relating to the issues:
    (i) ensure that the issues have been clearly defined;
    (ii) list those issues in a logical sequence;

(b) relating to the format of the present award:
    (i) decide the status of this particular award (is it to deal with all or only some issues, and if not all, which issue or issues?)
    (ii) check whether any particular format is required by party agreement or applicable rules;
    (iii) prepare (or if already prepared, check) a list of contents for this award;
    (iv) make a separate note of all (if any) matters to be decided in a future award;

(c) relating to decision-making on the substantive issue(s):
    (i) set down the basis (or bases if they differ as between issues) of decision-making;
    (ii) list, issue by issue (for those to be determined by this award), the questions or subsidiary matters for decision, the accumulated answers to which will generate his decision on that issue;
    (iii) collate (if not already done) the location of the individual submissions and the relevant evidence relating to each issue (and, if they are many or complex, do so separately for each question to be resolved within that issue);
    (iv) consider the submissions, the facts and the law (or other basis), question by question, then issue by issue, and reach a conclusion on each issue;
    (v) make his overall determination on the substantive issues by amalgamating those decisions, and, if appropriate, determining the remedy;

(d) and then, relating to further matters influenced by that determination:
    (i) consider and deal with any Value Added Tax implications;
    (ii) consider and make a decision upon interest if applicable;
    (iii) consider and decide liability for costs of the arbitration (and either determine the amount of recoverable costs or, more commonly, reserve that function to a future award);

(e)   and finally:
    (i)   condense that reasoning into a form suitable for inclusion as reasons in the award.

4.3   The components of decision-making

In the context of the preparation of an award, decision-making is broadly speaking of two kinds: deciding what needs to be recited, and resolving the matters in issue.

This exploration of the decision-making process intentionally addresses only the substantive issues, VAT considerations, interest and costs (Sections D–G), with a brief note on reserved matters (Section I). Other sections, along with these sections, are the subject of exploratory narrative in the illustrations starting in Chapter 5. For ease of cross-reference to the checklist in Chapter 3, and to the narrative and illustrations in later chapters, the relevant check-list sections are indicated, e.g. Section D.

Section D

(Parts of Sections B3 and C (hearing, submissions, evidence) might also apply.)

4.3.1   The substantive issue(s)

4.3.1.1   Basis of determination
In arbitrations in this country, by far the most common basis of determining the substantive issue(s) is English law. However, before starting the decision-making process the arbitrator needs to check that such is the case.

Under section 46(1)(b), that basis can be:

(a)   English law; *or*
    some other law, (if so, which?); *or*
(b)   'other considerations', e.g. justice and fairness (or similar definition); *or*
    trade/commercial practice; *or*
(c)   a combination of the two.

4.3.1.2   Collation of information
The purpose of careful collation is to produce a ready and speedy means of locating the information for decision-making, *issue by issue* (and for complex matters, sub-issue (or question) by sub-issue). That information is to be found in a combination of the pleadings (or other written definition of the issues, such as statements of case), the advocates' opening and closing submissions (either their written texts or

the arbitrator's notes of oral submissions), the documentary evidence, proofs and reports, and the arbitrator's notes of oral evidence and any relevant procedural matters which might have arisen at the hearing. Particularly if a hearing is of substantial length, the arbitrator can, possibly should, prepare a nightly precis of (or highlight) important evidence, notes or reminders of points still to be addressed, and so on.

Appendix 2 shows one style of form for manual use in collating that information. It identifies, party by party:

(a) the relevant issue;
(b) the location and nature of the advocate's various submissions;
(c) the evidence book and page numbers for oral evidence, indicating whether in chief, cross-or re-examination; *and*
(d) identification of and the page/item references of relevant documents.

Arbitrators adept at computer use can achieve even more by the use of software which can sort that information in various formats.

4.3.1.3 Consideration of submissions

Having drafted a logical sequence of decision-making, and having brought up-to-date and set down for compact ready-reference the basis of information retrieval just described, the arbitrator can now analyse and reason through each issue or sub-issue in order to reach individual, then composite, conclusions. Note, however, that when drafting the award it is reasons, not reasoning, which are to be there incorporated. (See 1.3.1.4 and 4.3.1.7 for the distinction.) The same quality of reasoning in reaching a decision is necessary whether an award is to set out those reasons or not.

Some issues are matters of fact, to be resolved solely on factual evidence, involving no contested matters of law. Some issues are matters of law[1], to be resolved solely on legal analysis, involving no contested matters of fact.

Some issues (the majority) are matters of mixed fact and law[1], to be resolved by determining ('finding') the relevant facts, determining (sometimes referred to as 'holding') the applicable legal principle(s), then applying such legal principle(s) to the facts found.

For each issue in the claim (and any counterclaim), the arbitrator has to decide between opposing contentions of fact and law (or other basis of determination). That applies first to liability, then to any 'quantum', i.e. the financial consequence. The information, i.e. the

---

[1] Where a basis of determination under section 46(1)(b) applies, that basis (equity, established trade practice, etc.) replaces these references to 'law' or 'legal principles'. This matter is explored briefly later in this chapter – see (7), as expanded at (7)(b) below, and see 9.3.7.1.

submissions and evidence, needs to be considered in 'bite-sized' pieces, i.e. logically-sequenced individual questions, the arbitrator's answers to which determine his decision on a single issue. This is referred to again under 4.3.1.4–6 later in this chapter.

The information to which the arbitrator must have regard in answering those questions consists of:

(1) agreed facts;
(2) uncontested facts;
(3) documentary evidence;
(4) inspection of the subject-matter;
(5) expert opinion evidence;
(6) facts to be found from consideration of evidence;
(7) applicable law (or, if appropriate, other basis of determination);
(8) the pleadings/opening and closing submissions.

These elements are considered below.

(1) Agreed facts
Whilst the arbitrator takes any agreed (or admitted) facts as though they had been proved and adopts them in his reasoning (or directly in the award if appropriate), no evidence is needed in order to prove them other than confirmation of them having been agreed or admitted.

(2) Uncontested facts
It can arise that some facts alleged by one party have been neither admitted nor denied by the other party. In that case, and if formally pleaded, such facts are taken as admitted. Where parties have no experienced representation, however, such admission should not be assumed unless the consequence has been spelled out to them at the outset. These facts can then be used in a similar way to agreed facts.

(3) Documentary evidence
Some documents will be agreed as being what they purport to be, without admitting the validity of information or statements in them or the interpretation of such information or statements. Some might be agreed both as to identity and validity or interpretation.

Documents accepted as such by both parties will usually be contained in an 'agreed bundle'. The parties should have clarified the status of the agreed bundle, usually that the documents therein are accepted solely as being what they purport to be. Documents which are not agreed as such will form part of a claimant's or respondent's individual bundle – and any of those which are to be relied upon will need to be proved, i.e. shown that they are what they purport to be.

The arbitrator should have checked on whether or not he was required to read any documents contained in the bundle(s) (and if so which) and whether he should make use of them all in his deliberations, or only those to which reference is specifically made by the parties. If he becomes aware of something in a document which would be likely to affect his decision but has not been aired in front of him, he should give both sides an opportunity of addressing him on the matter.

(4) Inspection of the subject matter
The arbitrator should inspect only in the presence of representatives of both parties, or of none but with the knowledge of both parties.

The purpose of any inspection should have been pre-defined and agreed. Is it simply so that the arbitrator can understand the submissions and evidence, and put matters into context? Or is it something more? Is it part of the submission of evidence, particularly in a quality arbitration? Is the arbitrator specifically authorised and expected in respect of the subject matter of the inspection to use his own specialist expertise in deciding between the opposing contentions? If so, is he required to make his decisions known, and receive further submissions on them, before finalising them?

The arbitrator should have made notes at the inspection, including who was present, what he saw and any decision(s) he had made (e.g. of quality).

(5) Expert evidence
In reaching a conclusion on the evidence of any expert, but particularly that of party-appointed experts, the arbitrator will have regard to:

(a) the expert's apparent standing in the relevant profession or calling, his or her relevant qualifications and relevant experience;
(b) the degree of impartiality exhibited in the expert's report and oral testimony, particularly under cross-examination; the extent to which, if that seemed appropriate, he or she was prepared to support the other side's contentions on any matter;
(c) the extent to which that expert had sought to find areas of common ground with the other party's expert;
(d) the facts or other information used as the basis of the respective experts' consideration and opinion, for these could differ, possibly explaining apparent differences of opinion;
(e) the extent to which the expert was prepared to change his or her opinion in the face of developments during the hearing, possibly relating to the emergence of facts differing from those upon which the expert had been instructed.

In the situation of 'opposing', party-appointed, experts, the arbitrator has to decide which version he prefers, having considered the foregoing and any other relevant factors. If he has appropriate expertise himself and has a specialist opinion of his own which differs from that of the experts, he must not utilise that expertise to 'oust' the expert evidence and replace it with his own. He must notify the parties of his dilemma, make them aware of his view, and allow them to address him on it – with time for them to consult their own experts.

In some instances, the 'opposing' experts will have produced a joint agreed report, either on all issues referred to them or on some of them. Such a report simply confirms that their respective opinions coincide on those matters, or for example that they agree what was shown by or at a joint inspection of, or testing of, some entity which is in issue. That agreement, however, is only one of the elements to be weighed by the arbitrator in deciding that issue or sub-issue.

(6) Facts to be found from consideration of evidence
Unless the parties have agreed that they want an unreasoned award, the arbitrator must give his reasons for his findings of fact. He must deal with the matter judiciously.

The burden of proof lies on the person asserting the affirmative of the matter in question. The normal rule in civil matters is that the degree of proof is that of the balance of probabilities. If a party making an assertion fails to convince the arbitrator that on the balance of probabilities the assertion is correct, that assertion is not proved. Proof is thus a positive function. However, the opposing party may (not must) adduce evidence to rebut, and thus negate, an assertion.

The arbitrator has to consider what weight he should give to individual items of evidence, or to the evidence of individual witnesses, whether in part or as a whole. Some evidence is simply not admissible or not relevant and is disregarded unless the parties have agreed that the strict rules of evidence are not to apply. Some is simply unconvincing or is too remote from the personal knowledge or recollection of the person giving it, thus influencing the weight to be given to it.

All other things being equal, consistent evidence of a witness will usually be preferred to that from one whose evidence is inconsistent from one part to another or with uncontested contemporary documents.

In some instances, expert (opinion) evidence as to whether asserted facts are likely, or possible, or impossible, can be of assistance. Additionally, an expert witness might on occasion give evidence of facts of which he is personally aware, as contrasted to alleged facts on which he has been instructed. By way of example, he might give evidence that he had seen a specific ship in a particular dock on a specified day, or that he had witnessed the destructive testing of a component relating to the dispute.

Issue by issue, having taken account of all the evidence in respect of both claim and (if there is one) counterclaim, the arbitrator has to consider whether or not a party has adduced sufficient, and sufficiently compelling, evidence to prove part or all of its contentions – and in either case whether or not the other party has adduced sufficiently convincing evidence of rebuttal to negate part or all of its opponent's case on the factual evidence.

(7) Applicable law (or, if appropriate, other basis of determination)
(a) General

As with facts, the arbitrator must, unless the parties require an unreasoned award, give his reasons for adopting (within the parameters of what has been submitted to him) a particular interpretation of the law. Where the parties are legally represented, those representatives will have cited such legal authorities (be they law reports or published authoritative texts) upon which that party relies. In some instances each will have cited different authorities, restricted to those thought to support its case. In others both will have cited the same cases but quoting different parts or putting different constructions upon them. The arbitrator must have proper regard to all of the cases cited, but must decide their relevance to the case he is dealing with.

In doing so he must check:

- whether the authority is of general application, or relates only to specific situations (such as a particular form of contract, or particular background facts);
  - If the authority is not of general application, the arbitrator needs to consider whether its specific factual/legal matrix was such (or had sufficient factors in common with the present case) as should nonetheless influence his decision.

- whether, if the authority is a judgment, he has been provided with a full text or only the part which seems to support the party citing it;
  - He needs to see the extract in its full context. That can sometimes change its apparent applicability. In some instances the extract is but part of the judge's reasoning which is ultimately discarded by the judge in reaching a different conclusion.

- whether the facts upon which a judgement was based are sufficiently 'on all fours' with the facts in the case he is deciding;
  - If the facts are sufficiently different the arbitrator has to consider whether or not the principles to be derived from the judgement are, or are not, applicable.

- whether the authority has binding effect, or is merely persuasive;
  - If it is only part of a judgement, did the point there relied on form an essential part of that decision (*ratio decidendi*) and thus be a binding precedent in sufficiently similar cases, or was it in effect an aside (*obiter dictum*), not being an essential part of the decision and thus have persuasive force only? Is it a judgment

of a court of less status than the High Court and perhaps has an element of persuasive force only?

- if conflicting authorities are cited, which (if any) is the closer to the facts and the issue in this case.

He must then bring the combined answers to those points to bear in reaching a conclusion.

(b) 'Equity' – or established trade practice

What of the situation where the basis of decision is to be 'other considerations' under section 46(1)(b)? If specific criteria have been established, they must be followed by the arbitrator. If authorities are cited to him he must, to the extent that they can be applied, have proper regard to the questions and criteria set out above regarding legal authorities, but adjusted to suit the circumstances. If they are from published articles or text books he will need information on the standing of the author.

If the basis for decision-making is 'in accordance with fairness and natural justice' or similar wording within the scope of section 46(1)(b), the arbitrator has to apply his judgement of what is in accordance with those criteria, rather than what is in accordance with strict legal principles.

If the basis for decision-making on the issues, or some of them, is established trade practice and:

- if the arbitrator is expert in the practices of that trade or calling; *and*
- if he has been appointed because of that expertise, and particularly if he has specifically been authorised by the parties to use that expertise,

he will need to have less detailed submissions on that practice than would an arbitrator without that expertise. Nonetheless, he should again avoid giving himself 'own evidence' – that is he must not use special or specialist knowledge of his own without making the parties aware of it and giving them an opportunity to address him on it (unless the parties have by agreement excluded that facility, e.g. possibly in a quality dispute).

An arbitrator without that expertise will be likely to need more detailed submissions on the relevant trade practice, possibly supported by evidence. Additionally, of his own volition or at the joint request of the parties, he might appoint an expert to report or a technical assessor to advise – but if that is done the requirements of section 37(1)(b) must be followed, after which the decision on the issues must be the arbitrator's own.

(c)   Unrepresented parties
If the parties are not legally represented, or not represented by persons with sufficient legal knowledge, the arbitrator still has to follow generally-accepted principles of law or of whatever the basis of determination is to be in the particular case. Whilst it is not for him to act in some kind of advocacy role, he must at least make known to such parties the general principles by which he will be guided, giving them an opportunity to address the point. In reaching his conclusion, he must not act in a manner or use knowledge or information which would take by surprise any reasonable party reading the award.

(8)   The pleadings/opening and closing submissions
The pleadings or the like, together with opening submissions, define the issues (subject to any subsequent allowable changes). The award must decide on *all* issues, no more and no less, subject to the possibility of a particular award stating that it deals only with certain issues, the remainder to be dealt with by one or more future awards.

The closing submissions are intended to pull the evidence and arguments into a cohesive form. They can and sometimes do list the questions which the arbitrator needs to answer in arriving at his individual decisions and his overall conclusions. Sometimes the advocates, or the parties, have agreed what those questions are. Sometimes the respective lists differ. In that case the arbitrator should have sought to resolve those differences at that time in order that there is no doubt as to what issues he is expected to determine; otherwise he must at least give the parties' representatives the opportunity to address him on the matter.

4.3.1.4   Determination, issue by issue
Being in possession of those elements of information outlined at (1) to (8) above to the extent that they are relevant, the arbitrator must for each issue of claim and counterclaim settle a logical sequence in which to consider and decide the individual questions (sub-issues) arising within that issue, and then to apply those decisions so as to resolve that issue.

By way of example (and for this purpose assuming no statutory provisions apply), say a decision on an issue depends upon:

(a)   whether a defect exists in a supplied product;
(b)   whether, if so, it constitutes a breach of the contractual specification for the product;
(c)   whether replacement under the product guarantee was conditional upon notice being given in a specified manner within a specified time; *and*
(d)   whether such notice was given.

How might the consideration of those elements be best sequenced? Unless the circumstances warranted otherwise:

(c/d)   If he were to find that replacement was conditional on proper notice (actual or constructive) and that it was *not* given, there would be no purpose in considering the other two questions, so the matter of notice could to advantage be considered first.

(a)   If he were to find that no defect existed, there would be no purpose in considering the matter of breach of specification, so existence (or not) of a defect could be considered second.

(b)   That would leave the matter of specification, *if still relevant*, until last.

In consequence, a logical sequence, and nature, of consideration for this particular issue would be:

● *Proper notice: was it necessary? was it given?*
This might simply be matters of fact to be proved, but if the requirement for or the manner or time of notice was contested it would be likely to become one of fact and law, particularly if the recipient of the contested notice had allegedly acted on it in some manner.

● *Defect?*
This might be a simple matter of incontrovertible fact, or might be a matter of fact to be proved, possibly supported by expert evidence of opinion, but there could also be legal contentions on, say, what constitutes a defect in general.

● *Breach of specification?*
This might be further sub-divided into:
○ *Proper interpretation of the specification*
This would be likely to be a mixture of expert evidence of opinion and legal contentions.
○ *Compliance?*
This could be a matter of fact, if it was apparent on the face (e.g. a permanently fixed component which was specified as to be adjustable). Otherwise it could again be a mixture of fact, expert evidence of opinion and legal contentions (e.g. whether the alleged non-compliance was *de minimis*, i.e. too trifling to be considered).

If instead of replacement by the respondent there was a monetary claim for repair or the cost of replacement, unless the amount as such was agreed ('figures agreed as figures'), there would be the further issue of whether or not the amount claimed was appropriate. This would usually be a matter of fact, possibly also involving expert evidence of opinion (e.g. whether it was a market price), and it is possible that matters of law could arise (e.g. mitigation of loss).

4.3.1.5 Accumulation of related issues (claim or counterclaim)
For the moment ignoring interest and costs, both of which are referred
to later, in a monetary award, or those parts of a mixed award which
are monetary, the amount to be awarded will often be derived from
the accumulation of the decisions on a number of separate issues. Each
of those might well in turn be the accumulation of the decisions in a
number of sub-issues. The manner in which those figures are summarised to arrive at a figure (or separate figures for claim and counterclaim) will vary, so that:

(a) If there is no counterclaim, the amount found to be due under the
claim is awarded (constituting a 'single event' when later considering costs).

(b) If there is a claim and a true counterclaim (i.e. one which is
not properly a set-off – see later at 4.3.3.1(3) and 4.3.4.1(5)),
the amounts found to be due under the claim and under the
counterclaim are awarded separately (constituting two events
when later considering costs). That is so even if, for convenience,
a single net sum is then shown (but a caveat is appropriate, for
the claim and counterclaim will almost certainly differ in terms of
any interest to be added, so that 'netting' can only be done after
interest has been dealt with; see later in this chapter under 'Interest' at 4.3.3);

(c) If there is a claim and a counterclaim which is entirely by way
of set-off, the amount found to be due under that counterclaim
is deducted from that found to be due under the claim, up to
the amount of the successful claim, giving a net amount
awarded (constituting a single event when later considering
costs).

In a non-monetary award, the decision might well similarly depend
upon decisions on several sub-issues. An example (the sole issue being
whether a binding contract existed) could be that under a series of
subsidiary issues the decisions were that:

- Mr A was authorised to bind that party in contract;
- he did in fact have relevant negotiations;
- he did reach agreement with the acknowledged authorised representative of the other party;
- the agreed terms had all the attributes necessary for a contract;

so the accumulation of those decisions leads to a decision on the issue,
i.e. a declaratory award that there was a binding contract between the
parties.

Whether monetary or otherwise, the remedy awarded must be
within the constraints of what has been pleaded. If a counterclaim is
pleaded as a set-off, the arbitrator would usually treat it as such. If the

arbitrator considers that a counterclaim whilst not so pleaded does in reality constitute a set-off, he should ask for submissions on that point before deciding.

4.3.1.6 Decision-making tasks relating to the substantive issues

The following is simply by way of a general reminder and an indication of possible tasks, to be read in conjunction with the preceding parts of this chapter (and adapted as necessary to give effect to any agreement by the parties on procedures).

(a) Generally:
- Collate the locations of evidence and submissions, issue by issue, if not already done.
- Decide a logical sequence for the consideration of the issues as such.

(b) Then, issue by issue in turn:
- Decide a logical sequence for consideration of the submissions and evidence relating to the constituent questions or sub-issues.
- Consider, within the confines of what is pleaded, the conflicting submissions and evidence and reach a conclusion on what relevant facts have been proved (or admitted).

(c) Then, if appropriate to the issue:
- Consider the conflicting submissions on the applicable law (if relevant) and reach a conclusion on the legal principles to be applied (and similarly with 'other considerations' under section 46(1)(b)).

(d) Then:
- Apply those principles to the facts 'found' and reach a conclusion on that issue, on liability and (if applicable) quantum (i.e. amount).

(e) Then, having reached conclusions on every issue to be dealt with in the current award:
- Accumulate and summarise the decisions on the various issues so as to produce the figure or figures, or declaration etc. to be awarded.

(f) and, if applicable:
- take account of any provisional order(s) issued under section 39.

4.3.1.7   Condensing reasons into a form suitable for inclusion in the award

What has been discussed in Section D, and is discussed in Sections E–G which follow, constitutes the process of reasoning. It is entirely inappropriate to incorporate in the award great tracts of that reasoning on a blow-by-blow basis. The arbitrator's reasons are, in effect, a summary or abstract of that reasoning, with recited detail being restricted to that necessary to make the stated reasons comprehensible. Some arbitrators set out reasons in a separately-headed section; some (perhaps most) prefer them simply to be apparent from their analysis of submission and evidence. The options are explored further in the illustrations in later chapters.

Section E

4.3.2   Value added tax

4.3.2.1   Background

Depending upon their area of activity, some arbitrators can publish many awards without ever encountering any problem concerning value added tax (VAT). Others meet the matter frequently and are familiar with its ramifications. Newcomers to award-writing, and perhaps some of those arbitrators in that first category, need to be sufficiently aware as to be forewarned of what can arise. The purpose of these brief notes, by way of introduction to and explanation of the short list of decision-making tasks which follow, is not to give some kind of potted guide to VAT; it is solely to make the reader slightly more aware of what might be encountered on that topic in an arbitration, and of the potential pitfalls.

(1)   General

These notes do not relate to the VAT component of the arbitrator's own charges.

VAT impacts differently, and to differing degrees, on various areas of commercial activities. Many arbitrators or potential arbitrators might have little in-depth knowledge of the ramifications of VAT in that wider perspective. An arbitrator should be able to appreciate the possible interrelationship between what he might award as a principal sum and any VAT implications relating to it – at least sufficiently as to enable appropriate questions to be posed to the parties' representatives.

Two ways in which VAT is most likely to arise in an award are:

(a)   where a claim is for payment for work done or services supplied and where such work or goods are or are likely to be subject to VAT; *and*

(b) where a claim for damages (or part of such claim) consists of reimbursement of the cost of having remedial work carried out by some third party and where that third party's charges have been or will be subject to the addition of VAT.

These are considered in slightly more detail later in this chapter.

Some parties to an arbitration will be 'taxable persons' (i.e. registered for VAT purposes), some will not. Subject to various exceptions or restrictions, a taxable person must charge, and remit to Customs and Excise, VAT on 'taxable supplies' of goods and services provided to others. A taxable person can recover from Customs and Excise as 'input tax' VAT paid on taxable supplies of certain goods or services supplied to him by others. If a taxable person gives a quotation for goods or services which constitute taxable supplies, but fails to mention VAT in that quotation, it is deemed to be inclusive of VAT.

In considering VAT implications relating to an award, it is essential to appreciate that:

(a) whilst, *as between the parties*, the arbitrator can and often does by his award have an effect on the value of services provided by one party to another which might be subject to VAT, he has no power over what VAT is chargeable *as such* by one party to the other, nor over what VAT is paid by a party to HM Commissioners of Customs and Excise; *and*

(b) as with the substantive dispute, an arbitrator cannot decide issues which have not been submitted to him – so that if the impact of VAT as between the parties has not been submitted to him he can make no decision on it.

A prudent arbitrator ensures that unless such matters are clear from the pleadings and/or submissions, as soon as is appropriate and certainly before the end of the hearing, he asks the parties, 'In the event that I award a sum to be paid by either party to the other, how, if at all, do you wish me to deal with any VAT implications?' – or words to that effect.

If the answer from both is 'Nothing', the arbitrator would recite that in his award and make no reference to VAT in the operative part of the award. He might, however, add a rider indicating that whilst he had been requested to have no regard to VAT implications, for the avoidance of doubt the amount awarded is exclusive of any VAT which might be applicable.

If, instead, the arbitrator is asked to deal with VAT implications in his award, he should ask for the appropriate submissions to be incorporated in the pleadings and ask to be addressed specifically on that element. The manner in which VAT is then dealt with in the award would depend upon the degree of discernible and incontrovertible precision and/or agreement on the extent to which any sum awarded (or any part of it) would be subject to VAT. That could result in either:

(a) a monetary award embracing (and identifying) the VAT component; *or*

(b) a declaratory award or declaratory adjunct to a monetary award, stating that a substantive amount awarded was subject to the addition of the appropriate VAT under the VAT Regulations.

To put those possibilities adequately into context, it is necessary here briefly to expand upon VAT.

Some standard forms of contract (for instance in construction matters, the VAT Agreement appended to the Joint Contracts Tribunal Standard Form of Building Contract) make specific provision for the adjustment of VAT following a court judgment or an arbitral award. An arbitrator must have regard to the implications, if any, of such agreement to the extent that it might affect the content and wording of his award.

Some services are subject to VAT, some are not. Again taking construction as an example, some work (supply of services) executed by a 'taxable person' contractor is subject to VAT, some is not; some elements of such supply are exempt, some are zero-rated; some payments for 'taxable supplies' follow the issue of a 'VAT invoice', some (interim) payments precede such issue. The arbitrator should, however, avoid being blinkered by knowledge of specific VAT provisions pertaining to areas with which he is familiar, be they construction contracts or some of the 'special' VAT provisions in other areas of trade or commerce, for those provisions are not of universal application.

In some instances there is little doubt as to the amount of VAT applicable to the sum awarded. It might be a situation where all goods or services supplied by that taxable person to a customer of that type are standard-rated (or all zero-rated or exempt, as the case might be). In such case, and if the *applicability* and rate of VAT is not itself in contention, any applicable VAT can be pleaded and awarded on a direct monetary basis – simply as a component of a monetary award, applying the appropriate rate to the relevant amount awarded.

In other instances, due to one or a combination of 'special provisions' in the VAT Regulations, it might at the time of pleading and of the award be impossible to determine the monetary amount of that VAT. Examples of special provisions include those where in certain circumstances the tax point is the date of receipt of payment by the supplier, or where a supply is of a mixed nature in VAT terms (i.e. part taxable, part not). Arbitrators should, however, be keen to ensure that there is no lacuna in their awards. Silence on the matter of VAT when there will be (or are likely to be) VAT implications, could appear to be a lack of certainty, or even of completeness. That is a situation where, if pleaded, the declaratory award option on the VAT element (see (b) above) is useful (even if, for enforcement purposes, less positive than a cash amount). See linked illustrations at 5.3.8.1 and 5.3.11.6, and note at end of 9.3.8.2 regarding interest.

If for good reason neither of those procedures (monetary or declaratory) are available or appropriate when a claimant is claiming payment for goods or services which have been provided, an arbitrator could add, after the operative section of the award but before his signature, a comment to the effect that, for the avoidance of doubt, the amount, or an identified portion of it, is exclusive of VAT (see 9.3.8.1).

The foregoing comments assume that in the main the parties are professionally represented. Such is not always the case. In small arbitrations, one or both parties, if unrepresented, could be at a complete loss on these matters. In those circumstances the arbitrator needs to take a more proactive role in teasing out the necessary information (perhaps taking the initiative under section 34(2)(g)). In so doing he must avoid actual or perceived partiality, or acting as advocate. Failing all else and provided that it could be relevant, he would have to adopt the declaratory 'such VAT as is payable...' option.

(2) Likely circumstances

An arbitrator might have to consider many elements of known or possible VAT implications in a variety of situations. The following examples, in which 'claimant' includes 'counterclaimant' in similar circumstances, illustrate some of them:

(a) *A 'taxable supplier' claimant's claim is for payment for work or goods which constitute taxable supplies under the VAT Regulations*

If the VAT element is not pleaded, the parties' requirements should first be established. How, if at all, do they wish it to be dealt with? If the relevant VAT element is pleaded, the award should include that VAT to the extent that a sum awarded is agreed by the parties to relate to taxable supplies. If pleaded but there is uncertainty as to the extent to which the sum awarded relates to taxable supplies, the award should incorporate a declaration that the appropriate VAT amount is to be added.

(b) *Part of a claim is for damages by way of interest on money due for such taxable supplies, but unpaid – and where that interest as such does not constitute a taxable supply or form part of the value of a taxable supply*

Interest on late payment does not attract VAT, so there are no VAT implications relating to the award.

(c) *A claimant's claim is for reimbursement of expense incurred in the rectification by a third party of faulty work executed by the respondent, where the charge for such rectification has had, or will have, VAT properly added*

If the successful claimant is *not* a taxable person, and therefore unable to recover that VAT as input tax, or is for other reasons not entitled to recover it from Customs and Excise, any applicable VAT on that work of rectification is part of the cost incurred. In those circumstances, if the amount (or basis of computation) of the VAT is known and not in dispute as such, it should be added to or included in the amount awarded. If the amount of VAT is not known, or is itself in dispute, it should be dealt with by way of a declaratory component of the award.

If the successful claimant *is* a taxable person and if it can recover the VAT from Customs and Excise as input tax, that VAT should not be added to or included in the amount awarded – for the ultimate cost to such a party does not include VAT.

(d) *A claim is in tort for damages*

If for injury caused, say, by impact to a structure or other property of the claimant where the remedy claimed is the cost of rectification of the physical damage to that property – and thus akin to (c) above, the position is as (c) above.

(If possibly arising out of the same incident), for *general* damages for personal injury and suffering, there is no taxable supply, so that VAT implications do not arise.

Related complications which can arise include:

- Supplies (goods or services) might have been wrongly classified (by a taxable person supplier) as taxable, or alternatively as zero-rated or exempt, and so treated later in an award, but at some future date, following a VAT audit by Customs and Excise, that classification is changed.

- The post-award reference of a relevant VAT matter to a VAT appeals tribunal, which could result in a VAT tribunal decision altering or cancelling VAT previously deemed payable to Customs and Excise and of which account had been taken in an award.

- In the event of a compromise settlement in a dispute concerning payment for an earlier taxable supply, that settlement possibly being recorded in an agreed award, the amount of that compromise settlement is taxable to VAT *if, and to the extent that, it is consideration for the earlier taxable supply.*

- If, when a dispute is compromised, the respondent has not admitted that the alleged supply took place, the amount of the payment is not necessarily treated for VAT purposes as consideration for the alleged supply, but as an estimate of the true worth of the claim – and as such might be outside the scope of VAT.

These are amongst the balancing factors in deciding whether or not, in those instances where such procedure could be adopted, a declaratory award concerning VAT might be the most desirable.

4.3.2.2 Decision-making tasks relating to VAT

The primary purpose of the foregoing notes, being more expansive than those on, for instance, interest, is to make arbitrators more, if only broadly, aware of potential complexities and to indicate that they should raise the matter of VAT with those representing the parties. Some cases might be simple in this VAT context; some are not. In view of that, any list of tasks here can only be to:

(a) Inform oneself sufficiently on VAT implications as to appreciate parties' submissions on the point.
(b) If submissions are silent on the matter of VAT, seek clarification.

and only then:

(c) Decide which of the procedures outlined above, or some other procedure, is applicable.
(d) Decide whether any function regarding VAT reimbursement (or payment of VAT as an element of damages) should be reserved to a future award; and then:
(e) Act accordingly.

Section F

4.3.3 *Interest*

Within the confines of the subject matter of this book, what follows is a brief introduction to the decision-making tasks relating to interest. Further reading on the topic of interest is essential.

4.3.3.1 Background

(1) General

An award of interest can arise in a variety of ways (see also at (5) below regarding post-award interest):

(a) Subject to the parties agreeing otherwise, the arbitrator has a discretion under section 49 of the 1996 Act to award interest:

- from such dates as meet the justice of the case:
  - on all or part of the amount awarded, for any period up to the date of the award;

        ○   on an amount claimed in, and outstanding at the commencement of, the arbitration but paid before the award was made, for any period up to the date of payment;

        ○   on an amount payable in consequence of a declaratory award (and logically, it would seem, for any period up to the date of payment);

   • from the date of the award (or later date specified by the arbitrator) until payment:

        ○   on the outstanding amount of any award, including any award of interest under section 49(3) and any award as to costs;

in each instance being simple or compound interest and at such rates and with such rests (intervals for compounding) as meets the justice of the case in respect of each of those factors. The arbitrator must have regard to other legislation (e.g. the Late Payment of Commercial Debts (Interest) Act 1998), if applicable, relating to the right to and/or the rate of interest.

(b)   Some contracts make provision for the payment of interest on unpaid amounts ('contractual interest'), and such interest can be an issue, or part of an issue, in an arbitration (and see section 49(6) of the Act).

(c)   There is provision at common law for interest to be allowable in certain circumstances, and section 81 (as well as sub-section 49(6)) has the effect of retaining that provision to the extent that it is not inconsistent with the provisions of the Act. Interest claimed as special damages (see below) comes within this provision, and if such interest is awarded it commonly becomes *part of the substantive sum awarded*. It is only mentioned under this present heading for reasons of completeness.

(d)   If related to a monetary award (e.g. relating to the cost of repairs awarded as damages) there is an award of *reimbursement* of related VAT paid, either as a further element of such monetary award or as an incorporated declaratory award, interest under section 49 has also to be considered in respect of such VAT items. In the first of those situations such interest would be in money terms in the usual way, in the second by a further declaration (section 49(5)).

(2) Restraint on discretion

Where a contract (or applicable rules, or other applicable statute) lays down:

(a)   the circumstances where interest is payable; *and/or*

(b)   the type of interest (simple or compound); *and/or*

(c)   the rate or basis of computing the rate; *and/or*

(d)   the commencement date from which (or period for which) interest is payable; *and/or*

(e)   other factors or requirements relating to interest,

the arbitrator has *to that extent* no discretion and must adopt those various provisions in each case.

If (see (iii) above) interest is claimed as special damages, provided that:

- the party claiming such common law interest pleads that element of damage; *and*
- proves that it suffered such loss by being held out of its money; *and*
- that such loss was in the reasonable contemplation of the parties when entering into the contract; *and*
- that the loss resulted from the wrongdoing of the other party,

the claiming party has a legal right to such damages; it is not a matter for the arbitrator's discretion under section 49.

(3)   Set-off?

In a monetary award, or the monetary part of a composite award:

(a)   Where there is a defence of set-off, or where the arbitrator has decided that a cross-claim though described as a counterclaim is in reality a set-off, the arbitrator awards a *net amount* in respect of the substantive issue (i.e. the gross amount awarded, less the amount awarded as set-off). Any interest on the substantive sum awarded (other than interest as special damages) is applied to that net amount unless special circumstances warrant otherwise (e.g. if the counterclaimant has already paid for remedial work). A cross-claim evaluated as at the time of the award bears no interest.

(b)   Where set-off does not apply, the arbitrator would initially award separate sums (to the extent that either is successful) on the claim and counterclaim, then consider interest *on each of them separately*.

(4)   Possible complexities

Matters relating to interest can be complex. They might refer differentially to claim and counterclaim, or involve different periods or rates for different issues, or might be a combination of different types of interest, or run over periods of particularly volatile movements in interest rates, or involve an intervening cause which could generate a pause or void in the interest calculations. In those circumstances the parties might request (or the arbitrator suggest or decide) that submissions, and possibly evidence, on interest be reserved for future delivery and a further award, i.e. after an award on the relevant substantive issue(s) has been published. The pleaded basis might, for instance in such volatile circumstances, be '[number] percent over the Bank of England base rate current from time to time' or similar terminology. Such submissions would incorporate calculations and any objections to them respectively.

Published opinions vary as to whether or not interest *up to* the date of the award may be awarded under section 49(3)(a) on any award of costs. It is for the arbitrator, having regard to what has been pleaded, to the parties' submissions on the point, and (if any) the decided state of the law at the relevant time, to consider and decide.

If the arbitrator decides against awarding interest, claimed or not, he must not simply remain silent on the matter. He must so state, giving his reasons. Where the parties are not represented, or they are represented but interest has not been pleaded, the prudent arbitrator will raise the matter and invite submissions – then consider the matter and decide.

(5) *Post-award interest*

Regarding post-award interest (section 49(4)), the 'paying' party is aware, in advance of its possible default in payment, just what consequences will follow, for those consequences have been spelled out by the award of that interest in the award. The rate should still, however, be compensatory, not punitive (although some arbitrators apply a modest increase). Unless prior-agreed by the parties, it must have proper regard to the justice of the case and have regard to the Late Payment of Commercial Debts (Interest) Act 1998 *if applicable*, pleaded and the subject of submissions. It is obviously not possible to calculate the amount.

This post-award interest may be awarded on:

(a)   the outstanding amount of the award (it might not all be/have been paid at one time);

(b)   any interest awarded under section 49(3); *and*

(c)   any award as to costs.

4.3.3.2 Decision-making tasks relating to interest

Once again, what follows is simply by way of general reminder and an indication of possible tasks, to be read in conjunction with the preceding parts of this chapter.

This list is not necessarily exhaustive. To the extent that the functions listed are, or might be, applicable to the particular award (if monetary, or there is a monetary impact of a declaratory award), having reached his decision on the substantive issues, the arbitrator should deal with the following (and any other matters relevant to the particular case) *and act accordingly*, thus giving more detailed effect to the checklist:

(i) Clarify just which type of interest (if pleaded) is claimed and whether he has discretion or not.

(ii) Check whether party agreement, or the contract or rules if either are applicable, or any other statute under which the arbitration is brought or which embraces or applies to it, exclude or amend any of his powers relating to interest; give full regard to such exclusions and amendments.

(iii) Decide whether to deal with interest (if applicable) in the current award, or to reserve it to a future award.

If interest is to be dealt with in this award (or when later dealing with reserved interest), continue as follows, and act accordingly, with due regard to the matters discussed above under this heading:

(iv) Decide in principle whether it is appropriate to award interest and if so whether it is to be on the whole or part of the sum awarded.

(v) Check whether any counterclaim or cross-claim is in reality a set-off (or was so pleaded).

(vi) Check whether before the award was published there had been any payment(s) of part or the whole of any amount claimed in, and outstanding at the commencement of, the arbitration; if so, consider interest thereon, as well as on any remaining balance now awarded.

(vii) If interest has been claimed in respect of any of those situations where the award of interest is not discretionary:
- Decide whether all requisites for entitlement have been complied with.
- Decide the proper commencement date for the computation of interest and, if the rate is at his discretion, decide the rate (or different rates for different periods); (the matter of simple or compound interest is likely to be pre-determined in those non-discretionary situations).

(viii) Regarding interest under section 49 (i.e. discretionary unless the parties have otherwise agreed):
- Decide whether the circumstances are such that he should in the exercise of his discretion award such interest; if he does:
- Decide the commencement date in respect of interest on:
  o the substantive sum awarded, or part(s) thereof; and decide the date(s) up to which interest is to run (up to no later than the date of the award);
  o any amount which was claimed in, and outstanding at the commencement of, the arbitration and has since been paid; and decide the date up to which interest is

to run (up to the date of (each if more than one) such payment(s));

(ix) Decide the appropriate rate (or rates) to apply (which should normally be based upon commercial reality – unless the justice of the case indicates more or less), whether to be simple or compound and, if the latter, with what rests (commonly likely to be quarterly).

(x) Decide whether there are good grounds for ordering that the period for the computation of interest shall be non-continuous, specifying any void(s).

(xi) Make the calculation or calculations, and have it/them checked; both the basis and the amount will normally appear in the award – or, having determined the basis, have the parties provide calculations for the arbitrator's prior checking, amendment as necessary, and incorporation.

Showing both basis and amount permits any error to be recognised, and corrected under the 'slip' rule, provided that the parties have left the arbitrator with that power.

If interest awarded is to relate to the financial consequence of a declaratory award, that interest, depending on the circumstances, will not always be capable of calculation to show as a figure, unless reserved to a later monetary award.

Regarding post-award interest, the arbitrator's decisions will be restricted to:

(xii) Is the award to carry interest or not?

(xiii) If to carry interest:
- Is it to be on the substantive sum, and/or on any award of interest under section 49(3), and/or any award of costs?
- Is it to be from the date of the award or some later date (and if later, from what date)?
- Is there any other relevant statutory provision for interest on late payment?
- At what rate, or rates (or basis of rate); simple or compound; period of rests if compound?

More than one basis of interest might apply within a single award. For instance, contractual or common law interest (or interest as special damages) might run to a date prior to the award, with discretionary interest (if awarded) running therefrom on the total of the substantive sum plus that initial interest.

If (as in the default position in section 52) an award gives reasons, reasons should be given for decisions on interest as well as for the substantive award. However, as the awarding of interest has become very much standard practice, such reasons can usually be restricted to those why interest has *not* been awarded, or why interest is non-continuous, or why there are unusual provisions as to rate or breaks or period.

Section G

4.3.4   *Costs*

Within the confines of the subject matter of this book, what follows is a brief introduction to the decision-making tasks relating to costs. Further reading on the topic of costs is essential.

4.3.4.1   Background

These notes identify some of the principal matters to be addressed by the arbitrator when considering costs. The topic is much wider than this brief outline, so that decision-making must be based upon a much wider awareness of the topic.

(1)   General

So far as the recoverable costs of the arbitration are concerned, in arbitrations of any substance the award on the substantive matter will usually deal only with liability for costs. That leaves the amount (the quantum) of those costs, if not agreed between the parties, to be the subject of future submissions and a future award. The principal exceptions, where quantum can be dealt with at the same time as liability, are:

- very modest cases, where the amount of costs involved might be known and claimed at the hearing;
- costs under any institutional schemes of 'consumer' arbitration which have provisions that enable quantum of costs to be dealt with.

Determination of quantum, whether as a component of a composite award or as a separate function, is in effect an arbitration on recoverable costs. That determination follows the same principles as for a monetary award but within the constraints of section 63.

If, after the dispute in question has arisen, the parties agree that whatever the result of the arbitration a party is to pay 'the whole or any part of the costs of the arbitration' (see section 60), the arbitrator must award in compliance with that agreement. Determination of the amount in those circumstances will remain with the arbitrator (subject to section 63(4) and subject to the effect of any party agreement under section 63(1) as to what costs are recoverable).

If under section 65 the arbitrator has previously placed a limit on the amount of recoverable costs, his award on liability for costs should be qualified to the effect that such liability is limited to that extent. The provisions of section 63 would still apply, but constrained by that limit. See also under 'Quantum' at (6) below.

If a claim for costs had been presented to the arbitrator at the appropriate time but not dealt with in his award, a party can apply under section 57(3) for an additional award making good that omission, and if the arbitrator then fails to do so he would leave himself open to challenge under section 68(2)(d).

(2) *Costs follow the event, except...*

It is possible for the parties to agree that the arbitrator in determining liability for costs may (or shall) have regard to factors other than overall success. That might possibly be on an issue by issue or some other specified basis. In the absence of such agreement, the starting point is the 'loser pays' maxim, 'costs follow the event' – and section 61(2) is quite specific about that, but with the proviso 'except where it appears to the tribunal that in the circumstances this is not appropriate in relation to the whole or part of the costs'. In diverging from the basic position, when deciding upon liability for costs the arbitrator has to consider and take proper account of the following in respect of both claim and counterclaim:

(a)   Is there a valid agreement under section 60 as to which party shall pay the whole or any part of the costs?

(b)   Has he previously made any 'costs in any event' orders, penalising a party for some procedural disruption or the like? Does such an order 'award' costs to an otherwise unsuccessful party?

(c)   If so, has he (as should be the norm) confirmed those orders as part of his award?

(d)   Has the successful party been unduly extravagant in its conduct of the hearing or preparation for it, or unreasonable or obstructive in the general conduct of the arbitration?

(e)   Have substantial costs been incurred on an issue or issues on which the otherwise successful (overall) party failed?

(f)   Has a claim been withdrawn (and particularly if at a late stage), leaving only incidental minor matters to be determined?

(g)   Has there been unreasonable exaggeration of a claim?

(h)   Are there discrete matters which render it appropriate to make a proportionate award on recoverable costs?

(i)   Has the successful party failed to accept a properly-constituted offer of settlement and failed to achieve a better result by continuing (see (4) below)?

(j)   Are there other factors which it is proper to consider?

If any of the answers are in the affirmative he may either allow the successful party only a proportion of its costs, or additionally make it liable for part of the 'losing' party's costs. If the arbitrator departs from the position of costs following the event, he must give reasons for that departure.

Regarding factors affecting costs of claim and counterclaim see (5) below; as to the quantum of costs and the basis of determination, see (6) below.

The requirements of section 33 apply equally to the matter of dealing with costs. The arbitrator must not have regard to matters which have not been raised by the parties or to which he has not drawn their attention, giving an opportunity for them to address such matters.

When an award grants a non-monetary remedy, the 'event' might seem less clear cut. The same can apply with what are in this book referred to as 'supportive awards' (see Chapters 2 and 7). In many instances, however, that perceived problem is substantially illusory. Has a party been awarded the declaration, or order for performance, or injunction, or rectification which it sought? If the answer is 'yes', then that is the event – and the successful party should be granted its costs, subject to the usual exceptions. A party might be only partially successful, say by not being awarded the full declaration or the full extent of performance, or the like, that it had sought. In those circumstances, the general principles relating to costs still apply. If the 'shortfall' from what was claimed was relatively insignificant, then the award of costs would be unlikely to be affected. If, however, that shortfall was of more consequence, it is likely that some form of proportionate award of costs would be applicable.

(3) Basis of determination of recoverable costs

As for the determination of what are to be recoverable costs, the default position under section 63(5) is, in brief:

(a) a reasonable amount in respect of all costs reasonably incurred; *and*
(b) any doubt in respect of those two criteria to be resolved in favour of the paying party.

There can, however, be circumstances where it is appropriate that recoverable costs should be determined on a basis different from that in section 63(5). The parties might have agreed on a different basis under section 63(1). A party might have acted in such a way that the other party makes an application that, should it be awarded costs, they be determined on some higher, more stringent, basis. In that last instance, if, having heard both parties on that point, the arbitrator decides to award on such higher basis, he must specify in his award what that basis is to be, having given reasons for its adoption. Before considering such a basis, an inexperienced arbitrator should indulge in wider reading – on 'indemnity costs' and the like.

(4) *Effect of offer of settlement*

If there has been an offer of settlement of the substantive dispute or an identified part of it by a 'Calderbank letter' or a sealed offer, the arbitrator becomes aware of the offer and its rejection only after he has reached his decision on the substantive matter, and has issued an award but reserving costs. To be effective the offer must have been for a principal sum in settlement of the dispute, *plus* interest and costs up to the date of the offer, or a specified date shortly thereafter so as to allow time for consideration.

If such a properly-constituted offer has been rejected, and the amount subsequently awarded in respect of the subject matter of that offer *does not exceed* the amount of the rejected offer, the otherwise successful claimant will normally be awarded the relevant costs only up to the date when the offer could reasonably have been accepted, with the other party being awarded costs from that date on. In making that comparison, however, it is essential to compare like with like. For instance and for this comparative purpose only, the interest component must be calculated up to the same date in each case. This can be done by notionally recalculating the interest awarded, taking it up to the applicable date of the offer, i.e. as though it had been awarded on that date.

(5) *Set-off?*

As indicated earlier at 4.3.1.5, where a cross-claim is a true counter-claim and not a set-off, that can constitute a separate event in terms of costs liability. Thus the arbitrator in such a case can award to the claimant its costs of a successful claim, and to the respondent its costs of a successful counterclaim. In both instances the costs might be abated as discussed at (2) above. Some arbitrators seek to circumvent that degree of duplication by obtaining the parties' agreement to him making an overall proportionate award of costs to one party, taking account of the respective complexities and degree of success. What needs to be borne in mind in considering that option is that what the arbitrator sees at the hearing is only part of the picture relating to costs and the manner of or reason for their being incurred. Effective preparation by a party can dramatically reduce what needs to be adduced at the hearing, and vice-versa. If any streamlined procedure for dealing with costs in a claim and counterclaim situation is to be considered, it is best if that is done following an award on liability which, say, awards costs of the claim to the claimant and costs of the counterclaim to the respondent. Only then, and having reserved that function, would the arbitrator receive submissions on such amalgamation or not, preparatory to the determination of quantum (taxation).

As also indicated earlier, where there is a successful set-off, the claim and (i.e. less the) set-off are treated as one event for costs purposes.

(6) Quantum

Sometimes, for whatever reason, the arbitrator is not to determine and award the amount of recoverable costs (whether in this award or later). In that instance, unless the parties have agreed otherwise, it would be appropriate for him to indicate, in his award on liability for costs, the basis upon which the amount should be determined. That can be of assistance to the parties in seeking to agree quantum, or to the court if such determination is referred there. In accordance with section 63(5)), that basis would be a reasonable amount in respect of all costs reasonably incurred etc., except in unusual circumstances. What is reasonable is usually a matter to be resolved as part of that later function of determining quantum. If, however, the arbitrator is aware at the time of awarding on liability that certain unusual costs have or have not been reasonably incurred, he can qualify his liability award by reference to such costs (provided that no party agreement under section 63(1) prevents him from so doing).

Brief reference has been made at 4.3.4.1(1) to the possibility that the arbitrator has previously, under section 65, placed a limit on the amount of recoverable costs. That limit might have been in respect of the (whole) arbitration, or 'any part of the arbitral proceedings'. Any such order, and any variation of it, must have been made in advance of the relevant costs being incurred. A likely candidate for an order relating to part of the proceedings is the potentially expensive discovery process (i.e. the listing, production and inspection of documents). In ultimately determining quantum of recoverable costs, whether in the award on the substantive issues or, as more likely, subsequently, the arbitrator (if he is to perform that task) must have full regard to such overall or partial limitation order(s).

A party might under section 38(3) have been ordered to provide (and/or has provided) security for the costs of the arbitration. As well as reciting that as a fact (checklist B2(k)1), the final disposition of that security must be dealt with here and in the appropriate operative section G (i.e. in this or a later award depending upon circumstances).

Where counsel has been involved, and particularly if the parties have agreed that determination of the quantum of costs is *not* to be by the arbitrator, then provided that he considers such involvement to be warranted, the arbitrator normally (within the operative section G) endorses his award 'Fit for Counsel'. That acts as an indication to others, or a reminder to himself if he does subsequently determine that quantum, that the case warranted the involvement of counsel. Any consideration of endorsing '*Not* fit for Counsel' should be approached with considerable caution and only after checking whether counsel was necessarily or reasonably involved because of some factor of which the arbitrator is unaware.

See at 4.3.3.1(5) regarding post-award interest on an award of costs.

(7) Arbitrator's fees and expenses

When, in the operative part of the award, dealing with interparty liability for his own fees and expenses, the arbitrator normally shows the total of those fees and expenses. In determining that total, he must act judicially, in effect as though he were considering the quantum of a party's recoverable costs. In doing so he will clearly have regard to whatever, if any, terms have been agreed (usually on appointment) – and also his duty under section 33(1)(b). In this connection, also see sections 28 and 64 of the Act.

Problems can arise where the tribunal is other than a sole arbitrator, if each party has appointed an arbitrator and each one has been appointed on the basis of different fee terms. That is one of the matters which section 64 addresses, but as this book is based upon a sole arbitrator situation it is inappropriate to develop the matter further here.

4.3.4.2 Decision-making tasks relating to costs

Once again, what follows is simply by way of general reminder and an indication of possible tasks, to be read in conjunction with the preceding parts of this chapter.

This list is not necessarily exhaustive but, to the extent that items are or might be applicable to the particular arbitration, the arbitrator, having reached his decision on the substantive issues (and, if applicable, on interest) should deal with the following, thus giving more detailed effect to the checklist, **and act accordingly in determining how costs are to be dealt with**:

(i)   Check whether party agreement, or the contract or rules (if any of those are applicable), or any other statute under which the arbitration is brought, exclude or amend any of his powers relating to costs; give full regard to such exclusions and amendments.

(ii)   Decide whether to deal with costs, or liability for costs, in the current award, or to reserve the whole, or only quantum, to a further award.

If to be in this award (or when later dealing with costs), continue as follows and act accordingly, with due regard to the matters discussed above under this heading:

(iii)   Check whether any valid agreement on who is to pay costs exists under section 60.

(iv)   Check whether any counterclaim or cross-claim was in reality a set-off (or was so pleaded); and whether one event, or more.

(v)   Consider whether there are valid grounds why costs, or any part of costs, should not follow the event (see earlier); if so, decide basis.

(vi) Consider any application that costs should be determined on a different basis from section 63(5).

(vii) Check and list (for confirmation in award) any costs in any event orders previously issued and whether any further exceptions should apply.

(viii) Check whether any limit has been imposed under section 65.

(xi) Check whether any security for costs has been given under section 38(3).

(x) Check whether there has been any provisional order under section 39 relating to an interim payment on account of costs.

(xi) Check whether any valid offer of settlement (Calderbank letter or sealed offer – or on occasion an open offer) has been made and not accepted.

(xii) Consider whether to set any criteria as to recoverable costs – and consider whether to endorse 'Fit for Counsel'.

(xiii) Check whether the *quantum* of costs (and the matter of post-award interest on costs) can properly be determined by this award; if so, seek submissions and do so; if not, reserve that function (unless the parties agree otherwise).

(xiv) Consider how the decision on liability for costs is best to be reflected in deciding liability for the arbitrator's fees and expenses; consider whether interparty reimbursement (see 5.3.11.7) should carry interest (but see 4.3.3.1(4) paragraph 2).

Section I

4.3.5 *Reserved matters*

This relates to matters to be left over to be resolved by a subsequent award. Unless such reservation has been agreed between the parties, the arbitrator needs at some stage to decide what, if any, matters should be so reserved. There could be, for instance, remaining issues, interest, or costs; or less apparent matters such as the consequences of subsequent failure to comply with a performance award. He should specifically reserve those elements, commonly by an item following the operative part of the award and prior to his signature. He should also maintain a further checklist so as to ensure that no reserved matters are subsequently overlooked.

Finally

In all the arbitrator's decision-making, he must have full regard to those requirements of section 33 of the 1996 Act.

Part C
Developing the principles

5   Synthesis of an award

5 Synthesis of an award

5.1 Introduction to the illustrations

Comments and illustrations relate to English law and are on the basis that, unless stated otherwise, all 'default' provisions of the 1996 Act apply without alteration and that there are only two parties.

First-time arbitrators and those studying the subject usually have little difficulty in absorbing the general format and sequence for a basic award which is often illustrated. They should similarly have little difficulty with the expanded checklist shown in Chapter 3. It is in putting such a sequence or list into practice that some have difficulty. Headings or items which seemed abundantly clear at first reading become imagined minefields later.

For that reason, the items in that checklist are considered in this and succeeding chapters, briefly exploring matters to be dealt with under each item, with illustrations of award clauses.

All award content illustrations in this and succeeding chapters are shown shaded. The illustrations in this chapter, but merged together without the explanatory notes, effectively constitute the first of the group of 'substantive awards' – a monetary award. As it forms the underlying basis for the later illustrations, it is repeated as Appendix 3 without those notes. The notes, isolated from the illustration, also form a basis for, and should be read in conjunction with, illustrations relating to other types of award in later chapters. For ease of cross-reference, the sub-headings in the narrative show the relevant checklist reference, when used as a heading, in square brackets: e.g. [B1(a)], in addition to the decimal enumeration. Within this chapter only, those alpha-numeric references are complete and are in sequence. They are used for identifying other items in this chapter to which reference is made, and to enable those items readily to be considered in the context of the checklist as a whole.

Otherwise, square brackets in the various award illustrations indicate where something is to be inserted, usually showing the nature of the insertion, e.g. [*date*]. Words shown in braces {} are further illustrations of matters which might arise. Explanatory narrative relates to 'default' provisions (i.e. where the parties have not agreed otherwise), so might not apply, or might be varied, if the parties *have* agreed otherwise.

The illustrations are not intended as exemplars or advice. The content, sequence and form of words used in them are intended to be simply illustrative, to clarify the checklist descriptions – capable of being adopted and adapted, or of giving a basis from which each reader can construct a personal award content, sequence and style. Personal style, or the circumstances of each particular case, can generate different approaches.

The illustrations, including the synthesis in this chapter, adopt the same Sections A–J as the checklist. The same basic scenario is used throughout the illustrations, but is expanded in Chapters 6 and 7 to suit each subsequent award type or situation. Where parts of awards remain substantially unaltered in the succeeding illustrations in Chapters 6 to 8, they are not repeated. The principal additional or varied clauses only are shown or indicated. The explanatory narrative should be read in conjunction with Chapter 4.

In particular, and to avoid confusion as to which changes are generated solely by each succeeding *type* of award, the arbitration agreement and mode of appointment remain unchanged for the illustrations in Chapters 6 to 8, with a few necessary exceptions which will be clear from the context. Variants of content (contract or tort, different modes of appointment, and so on) are explored via the illustrations in Chapter 9, their position in the award being mainly located by checklist numbers.

Matters shown in the expanded checklist should appear in an award (and appear in the illustrations which follow) *only* to the extent that they are necessary for the proper understanding of, and validity of, that award. No 'nil report' is needed except where recording such things as there being no 'default' powers excluded by the parties, or similar situations. However, and so as to further explain the intent of the checklist items, each of those items is shown in this first illustration, with brief notes, whether the item applies to it or not. Thus this chapter incorporates an explanation of the checklist items, together with illustrations which cumulatively comprise an award.

5.2   The basic scenario

Techsolut (GB) plc (TGP) is a multi-specialism engineering company which designs and manufactures what it terms 'hi-tech equipment for the entertainment industry'. Sophisticated Theatres Ltd (STL) is a company which owns and operates several theatres which specialise in lavish performances.

STL had a recurring need for complex three-dimensional sets (units), for use in connection with orthodox 'flats' and the like. The units had to be mobile, both manually and by electric power. They had to have a variety of lighting and other effects incorporated, with provision for remote control. In addition, they had to be capable of a variety of configurations, and have provision for speedy attachment and removal of panels and components to replicate two-storey buildings and other structures, parts of ships, double-deck vehicles and so on – with easy access for performers from one level to another. As one or more units might on occasion have to be moved temporarily between theatres in the group, it was also a requirement that they must be readily demountable and transportable. However, that explanation does not impinge directly upon the first illustration in the synthesis below.

After some discussions, STL entered into an agreement in writing with TGP whereby TGP would design, manufacture and install two such units at STL's theatre in Weiringhampton, at a price of £195,752.00. The contract was evidenced by exchange of letters, with few terms other than technical matters, but it incorporated a brief arbitration agreement. The units were duly installed. During installation, TGP redecorated the permanent stage structure and related backstage areas, a type of work outside the company's expertise and, they contended, done solely as a favour to STL and on instruction as extra work.

TGP sought to charge for that work of redecoration. STL refused to pay, on the basis that TGP had at all times known that those areas were to be redecorated, and should have included that work in the contract price. Being unable to reconcile their differences, TGP and STL appointed Jonathan Fairley as sole arbitrator.

Further details will become clear from the narrative which is interspersed with the synthesis, and from the succeeding illustrations.

5.3   Synthesis of an award

Before beginning the drafting process, it is helpful to prepare an *edited* checklist and/or a proposed list of contents. In this illustration, those items with no indicative award text are the ones which would have been deleted in preparing such an edited list.

Reminder
In practice and so as to suit individual awards, it will often be the case that two or more of the detailed items in the checklist shown in Chapter 3 can and should be merged into composite award items, or

the sequence of items can to advantage be changed, or items moved from one of the indicated sections to another.

Section A

5.3.1   *The heading*   [A]

Linked with the heading, it is appropriate to consider the cover to an award, for it has a function other than simply that of protection. Particularly where there is a series of awards relating to a single dispute, the cover provides a basis of speedy identification, usually setting out only:

(a)   the heading and identification of the parties and the particular award, in the same form as indicated below for the heading in the award proper;
(b)   the date of the award; *and*
(c)   the name and address of the arbitrator.

Particularly if an award is of substantial length or complexity, or of unusual sequence, a list of contents can to advantage be inserted before the award proper.

Traditionally, if not essentially (and see 9.3.1 regarding alternatives to this stylised 'In the matter of . . .' wording), the heading will usually commence with the title of the relevant arbitration statute, in this instance:

In the matter of the Arbitration Act 1996

continuing, for this example, with:

and in the matter of an Arbitration

between

Techsolut (GB) plc   **Claimant**

and

Sophisticated Theatres Ltd   **Respondent**

followed by a heading indicating the nature of the award (First, Second, etc. Award; Final Award; Additional Award; Consent Award; etc.), e.g:

FIRST AWARD

If the arbitrator favours the option of initially setting out by way of preamble a brief historical outline, that may conveniently be placed here. If it paraphrases matters which are to be recited later, care should be taken to avoid tiresome repetition. The preamble can, if desired, incorporate the full identification of the parties, but if the intention is solely to paint a broad picture, leaving full identification and other details to be dealt with under 'Background', such a preamble might be:

> **Preamble**
>
> The dispute refers to work executed by the Claimant, a company specialising in the design and construction of theatre equipment, at Theatre STL at Weiringhampton on behalf of the Respondent company, which owns and operates a number of specialist theatres.
>
> The Claimant contracted to design, construct and install two complete 'hi-tech' mobile stage sets, including what were termed 'integral finishes'. In addition, they redecorated the permanent stage structure and related backstage areas, then sought to charge extra for that work, alleging that it had been ordered as an extra. The Respondent contended that the redecoration was, or should have been, included in the contract price and also contended that the charge for that work was too high.
>
> Being unable to compromise their differences, they have submitted the dispute to be resolved by arbitration.

Section B

5.3.2   *Background*   [B]

(or Introduction, Recitals, or whatever term is preferred and appropriate)

The preamble shown above can alternatively be inserted here.

Section B1

5.3.3   *Identification and jurisdiction*   [B1]

First, this (or other appropriate) sub-heading:

> **Identification and jurisdiction**

Here should be set out the non-contentious details of how and why the award has arisen, including the authority of the arbitrator. Sub-headings break up the text into recognisable and manageable components, so that particularly in a substantial award their use is likely to be beneficial. They might be derived from this illustration, or the arbitrator's own checklist, or be ad hoc to suit the circumstances.

5.3.3.1 Identification of the parties [B1(a)]

The identification can be dealt with in any logical, unambiguous and factually accurate manner. The method of doing so can depend, for instance, upon whether the dispute is in contract or in tort. This example is a contractual dispute. See 9.3.3.1 for example of a claim in tort, and 9.5 for brief reference to multi-party arbitrations.

If under section 52 of the 1996 Act the parties have agreed how they are to be identified in an award, the arbitrator must adopt that mode or precision of identification. Otherwise, opinions of arbitrators differ as to whether the full details of a corporate body should be given (such as registered address, possibly the operating address, and company number if a limited company), or simply its registered name (relying upon reference to the Registrar of Companies should there be any problem of identity). In favour of giving full details is precision of identity. Against that approach is the possibility, perhaps unknown to the arbitrator, of change of registered (or of operating) address or other details during the currency of the arbitration. If the address of a corporate body is quoted in the award it is not enough simply to set down that which applied at the beginning of the arbitration. The address at the time of the award needs to be ascertained and given, possibly described as 'of [*address*], previously of [*address*]'.

If a party is an individual, or a non-registered firm, the address should be recited so as positively to identify that party. Again, care must be exercised to ensure that it is correct and up to date at the time of the award.

The identification might be briefly set out:

> 1 The Claimant, Techsolut (GB) plc of [*address*], is a company specialising in the design and construction of theatre equipment; the Respondent, Sophisticated Theatres Ltd, of [*address*] owns and operates a number of specialist theatres.

Then, still under this heading of 'Recitals', or 'Background', or 'Introduction', or the like, should be shown the other background facts, such as:

5.3.3.2 Contract, or other relationship [B1(b)]

Here might be (in more, or less, detail – depending upon the extent of information recited in any preamble):

> 2 The Respondent engaged the Claimant to design and construct two hi-tech stage sets for use in and subsequent to the high profile re-opening after refurbishment (by others) of Theatre STL at Weiringhampton. The design, construction and installation of the sets were the subject of an agreement in writing, dated 5th November [*year*], evidenced by a letter of enquiry dated 1st October

> [*year*], a brief quotation dated 10th October [*year*], and a letter of acceptance dated 5th November [*year*].

This recital should continue with contract details, but only to the extent that they are relevant.

5.3.3.3   Law of the contract or other relationship   [B1(c)]

Unless there is a possibility that the law of the contract (or the law governing the relationship between the parties) could be other than English law, it should not be necessary to recite the applicable law. It does not arise in this illustration.

5.3.3.4   Arbitration agreement   [B1(d)]
5.3.3.5   Provisions for appointment   [B1(e)]

This is illustrative of a situation where two or more checklist items might conveniently be merged together.

It is necessary to make adequate reference to, or to recite, the arbitration agreement, whether that agreement was contained within some other agreement or was entered into when a dispute had arisen. Care should be taken to use the same terms as used in the agreement, e.g. 'arising under', or 'arising under or in connection with', etc.

Here might be:

> 3   The contract included an arbitration clause whereby in the event of any dispute arising under the contract such dispute was to be submitted to arbitration by an arbitrator to be agreed or, if not agreed within [*days*], to be appointed on the application of either party by the President or a Vice-President of [*name of body*] {The Theatre Industry Contracts Tribunal (TICT)}.

Any specifically relevant terms, including any incorporated 'party agreement' provisions on procedures or other matters would be recited here. In the present case there were none at that time. See 9.3.3.2 for appointments by 'appointing authorities' and 9.3.3.1 for 'submission agreements' (i.e. an agreement to submit and simultaneously submitting an existing dispute). See 9.4 for brief comment on tribunals other than a sole arbitrator.

5.3.3.6   Applicable rules   [B1(f)]

If it is likely that any institutional or other rules are to apply, this would usually have been provided within the arbitration agreement. However, it is not uncommon that an agreement or an appointment which does not at the outset provide for such rules to apply is the subject of post-appointment party agreement in writing to the effect that specified rules *shall* apply. If that occurs, that agreement should be recited (stating how and when made), identifying the rules and (as with all references to such rules) specifying any options or alternatives therein which have been chosen by the parties, together with any variations adopted by the parties (again in writing). Here the parties decided against using the industry's Arbitration Rules, thus:

> 4 Following correspondence between the parties' representatives the parties agreed, by letters of [*date*] and [*date*], that any such arbitration should *not* be subject to the Arbitration Rules issued by the TICT.

Whilst in this instance no rules applied, see Chapter 8, under 'Institutional awards'.

5.3.3.7   A dispute having arisen [B1(g)]

On occasion, there can at the outset have been problems in defining just what is in issue. If that had been the situation, it can be appropriate here simply to record that a dispute arose within the parameters of the arbitration agreement (mentioning any known counterclaim). However, this item can instead often be better combined with the item immediately following, outlining the dispute as such. That approach is adopted here.

5.3.3.8   Nature of the dispute and matters in issue [B1(h)]

The award should lead on to a brief initial description of the matters in dispute. The amount of detail to be shown, and its most logical position in the award, depends upon how precisely or not the parties have identified the dispute before either jointly requesting a chosen arbitrator to act or applying to an appointing authority to make an appointment.

If already crystallised in detail, that detail can to advantage be recited here or, if more logical in the particular circumstances, can be linked to the later entry regarding 'Appointment' (B1(i)).

If that degree of precision arises only on exchange of pleadings or statements of case, the detailed recital should be linked with, or follow, the later entry, 'Resultant party action' (B2(j)). In that instance the present entry could either simply cross-refer to the preceding preamble, or recite that text here instead.

In the present case, it is to be assumed that the parties had by the time of the joint invitation to the arbitrator already crystallised the matter, possibly giving:

> 5 A dispute has arisen within the parameters of the arbitration agreement, its nature and matters in issue being as follows:
>
> On completion of the contract works it was alleged by the Claimant that during the progress of the works of installing the sets:

(a) extra works were ordered (being backstage redecoration of surfaces not forming parts of the sets for which it was responsible and not being the type of work it normally undertook), without fixing any price or basis of pricing; and

(b) this caused difficulties for the Claimant, not least because the extra work was outside its province and expertise, even though its staff were multi-skilled, specialist technicians; and

(c) it valued the said extra work at £18,750 plus VAT, which amount it sought to recover from the Respondent; and

(d) the Respondent has refused to pay that or any sum in respect of the said extra work; and

(e) the Claimant now claims:

    (i) that sum of £18,750;

    (ii) Value Added Tax related to that sum;

    (iii) interest under section 49 of the Arbitration Act 1996; and

    (iv) costs

and the Respondent contends that:

(f) the said work was, or should have been, included in the Claimant's contract figure, which has been invoiced and paid;

(g) it did not at any time order the said work as an extra;

(h) even if, which it denies, it is found liable to pay for the said work the amount claimed is excessive; and

(i) the Respondent claims costs.

These matters are the subject of this arbitration.

If there were a counterclaim, that would be described here in similar manner. That is not done in this 'basic' award structure so as to avoid over-complication of options shown, but is addressed at 6.4. Similarly, recital of disputes and issues in the other types of awards listed in Chapter 2 are addressed in Chapters 6–9.

5.3.3.9   Appointment [B1(i)]

If it were not recited earlier, the award would then show how the arbitrator comes to have jurisdiction in the case. A simple reference to appointment might be:

6 Following a joint invitation signed by the parties and dated 2nd April [*year*] I, Jonathan Fairley MSc (Performing Arts), of Palladium House, Laburnum Way, Nevekirk, forwarded to both parties a form of appointment which, inter alia, set out my terms and conditions. That form was returned to me, signed by the parties on 9th and 10th April [*year*] respectively – and I signed it by way of

> acceptance of appointment on 12th April [*year*], thereby taking up jurisdiction in this reference.

If the terms of appointment incorporated special powers or obligations, or if such powers or obligations arose under the 1996 Act as 'default' provisions (i.e. not 'otherwise agreed'), or were later agreed 'in writing', those special powers or obligations should be mentioned here to the extent that they are *relevant* to this award. In the present case there are no unusual powers or procedures. (See the 1996 Act section 5 definition of 'in writing'.)

See 8.3.2 and 9.3.3.2(1) for appointment by a nominated appointing authority.

5.3.3.10 Challenge to jurisdiction [B1(j)]
In the event of any such challenge having been made it should be recited here, stating how it had been dealt with. In this case that does not arise, but see at 7.2.

5.3.3.11 Basis for decision [B1(k)]
Again if not previously dealt with and recited, if the parties at any stage agree that 'other considerations' (see section 46(1)(b)), or a law other than that of England and Wales, shall be the basis of the arbitrator's decisions, or any of them, that should be recited. That does not arise in this case, but see 9.3.3.3 and 9.3.7.1 for illustrations.

5.3.3.12 Other matters [B1(l)]
This checklist heading is a reminder to consider what, if any, further matters should be recited in the particular award.

Section B2

5.3.4 *Interlocutory procedural matters* [B2]

An award should not recite a blow-by-blow account of each and every interlocutory stage of the arbitration. Subject to that constraint, what follows here indicates matters to be considered. They will not always arise, they will often be better in a different sequence, and some groups of items might on occasion be combined to good effect.

First, a sub-heading:

> **Interlocutory procedural matters**

5.3.4.1 Transfers [B2(a)]
This checklist sub-heading is simply a reminder to adjust the checklist to suit the particular award.

5.3.4.2 Seat [B2(b)]

If not already recited as part of some other item, the juridical 'seat' should be identified unless there are no international implications and no possibility of enforcement being sought abroad. However, there is no harm in making the stating of the seat a standard practice in any event, unless the parties (under section 52) have agreed otherwise. Whenever and by whoever it was determined, some arbitrators prefer to state the seat at the outset, as one of the first few items in the award, some at the end with the substantive decisions, some linked with the stage at which it was determined (such as at a preliminary meeting). As its main purpose is to inform any court faced with an application on the award, there is much to be said for it being recited at or near the beginning, no matter when determined, or at the end, where anyone will look for decisions by the arbitrator. An example, whether positioned hereabouts or elsewhere, could amongst other possibilities be one of the following:

> 7   The parties have, by written agreement, designated the Seat of the arbitration as being England {and Wales}.

or, had those rules applied and had the TICT acted under powers under those rules:

> 7   The Seat of the arbitration has been designated by the institution vested by the parties with that power, vis the TICT, as being England {and Wales}.

Whilst neither the use of 'England' or 'England and Wales' is remotely likely to cause any problems or lack of particularity in practice, England and Wales share one jurisdiction and 'juridical' means, in this context, 'of or relating to the administration of justice'; and 'seat' means 'the place or centre in which something is located'. See 9.3.4.1 for further illustration.

5.3.4.3 Interlocutory agreement on rules, powers
or procedures [B2(c)]

Particularly if the arbitration agreement related to *future* disputes, and more so if coupled with an 'institutional' appointment, it is eminently possible that at this stage (between appointment and the preliminary meeting if such was held) there could be agreement on, or on further, procedures or arbitral powers. Any which are relevant to the award should be recited, either here or, if convenient and not confusing, by adding to those previously recited. In this instance there were none.

The types of matter likely to arise include agreement, or application, or order(s), as follows.

(1)   On procedural and evidential matters   [B2(c)1]
There is provision below (at B2(i)1) for reciting matters relating to a
hearing (whether there is to be a hearing as such or some other
procedure), and at B3(b) (later record of date and place where hearing
had been held). Inter-party (or inter-solicitor) correspondence might
well, however, generate agreement (or application for an Order) on,
for instance, the type of matters covered by sections 34, 37 or 38. They
can conveniently be recorded here, or later at B2(i).

(2)   On consolidation, or concurrent hearings   [B2(c)2]
The essential party agreement for these procedures (see section 35,
1996 Act) might arise now or later. The arbitrator has no such default
powers. If either of those processes is agreed at this stage it is likely to
affect a variety of subsequent procedural matters. That being so, that
recital of such agreement now can help explain later parts of the
award. Otherwise that agreement can be recited with one of the references to the hearing which occur later. (See 9.3.4.2(2).)

(3)   On appointment of experts, legal advisers or assessors   [B2(c)3]
This relates to tribunal (arbitrator)-appointed experts etc. under section 37. Whilst such appointments are relatively uncommon, again
they can influence subsequent procedural matters – and recital now
can help explain later parts of the award. Details of such appointments, their reason, the nature and extent of the briefs to the appointees, and other details, should be recited here or, if more appropriate,
later. (See 9.3.4.2(3).)

(4)   Relating to property the subject of the proceedings   [B2(c)4]
Relevant orders might relate to, for example, inspection, preservation,
sampling, and so on (see section 38(4)) and it is eminently possible that
the need for such an order could arise at this early stage. Again, whilst
on some occasions greatest clarity can be achieved if the fact of such an
order having been issued is recited here, on other occasions it can be
better recited later. (See 9.3.4.2(4).)

(5)   On preservation of evidence   [B2(c)5]
Once again, a party might at an early stage have been concerned that
important evidence in the custody or control of the other party could
be lost or destroyed. It might in consequence have applied for an order
under section 38(6). That can be recited here or later. (See 9.3.4.2(5).)

5.3.4.4   Use of 'uncommon' powers   [B2(d)]
The fact that an arbitrator has been granted uncommon powers does
not of necessity mean that those powers will have been used, or have
been used in respect of all issues. If it is necessary so as to explain or

give validity to an award, the method and extent of use of such powers should be recited. If, for instance, the arbitrator had under section 34(2)(g) taken 'the initiative in ascertaining the facts and the law', it should be recited in what manner and in respect of what matters that had been done – and how the parties had been kept informed on the manner in which that was being done. None apply in this intentionally simple case, but see 9.3.4.3 for further illustration.

5.3.4.5   Clarification or amendment of matters in issue   [B2(e)]
Sometimes the precise nature of the issues has needed to be clarified. Parties might amend what was, or appeared to be, in issue between them regarding the claim and/or counterclaim. In either event that should be recited. In this simple illustration there was nothing to recite here.

5.3.4.6   Previous awards   [B2(f)]
Had this award been one of a series where various issues were decided by separate awards, and had one or more such awards been issued, that fact would be recorded here, stating succinctly which issues had been decided by those awards. In such an instance the underlying recitals in the second (and any subsequent) award would usually be incorporated by reference, rather than being repeated in full. Any additional matters of recital arising since the previous award (e.g. agreed change of procedures) would however be set out normally. That does not arise in this illustration, but see the illustrations at 7.4.

5.3.4.7   Issue(s) to be determined by this award   [B2(g)]
Sometimes all issues are to be resolved in a single award. Sometimes, as just indicated, individual issues or groups of issues are to be resolved in a series of awards. Sometimes the parties or their advisers have agreed a specific list of issues to be decided; if so, that is likely to be in a form which can be repeated here. Sometimes the issues are pleaded in terms which, similarly, can be repeated or incorporated by reference to those pleaded terms. Sometimes the arbitrator has needed to deduce just what the underlying issues are, and in that case is likely to have submitted such a list to the parties for confirmation or comment. In this present case, there being no counterclaim, the item might be:

8   The issue to be determined by my award is: *Is the Claimant entitled to additional payment for executing the work in dispute and if so how much?* The parties have agreed that the sub-issues to be resolved in determining that issue are as follows and that I am to award such amount, if any, as I so determine:

(a)   was the disputed work included in the Respondent's letter of enquiry dated 1st October [*year*]?

> (b) if not, was it specifically included in the Claimant's quotation dated 10th October [*year*]?
>
> (c) if neither, did the Respondent subsequently order that work {it having been agreed that no issue is taken as to whether or not the Respondent was entitled to issue such order}?
>
> (d) if so, did the said work constitute either extra work or a collateral or other contract (but for this purpose without any need to decide which) entitling the Claimant to extra payment?
>
> (e) if so, what amount was a fair charge for that work?

The phrase in braces {} has been inserted here simply to limit the extent of the present illustration.

5.3.4.8 Meetings: preliminary, further, pre-hearing review   [B2(h)]

At this stage, it can be appropriate to recite simply the fact of such meeting(s) (or their equivalent by conference call or other means) having been held, leaving to a later section any reference to what transpired, but *to the extent only that such recital is necessary so as to explain or validate the award.* In relatively small or simple arbitrations, reference to such meetings, any agreements made then, any orders issued as a result of the meeting and, to the extent necessary or helpful, details of compliance, can all be combined in one item. In the present case, reference to a preliminary meeting and resultant orders are combined, compliance being left for another item. See notes under B2(i) below, and 9.3.4.4, for further exploration.

An example could be:

> 9   I conducted an initial preliminary meeting on 19th April [*year*] at my office at which the parties were represented by:
>
>> Claimant   :   Mr GJ Wyttle, Technical Director;
>> Respondent :   Mr Justin Blackwelle, Counsel,
>
> following which I issued my Order No 1 giving directions for the conduct of the reference, providing for:
>
>> an attended Hearing;
>> exchange of Statements of Case;
>> limited disclosure and production of documents;
>> exchange of witness proofs and experts' reports; and
>> meetings of experts.
>
> I issued further orders from time to time, two of them with a sanction as to costs (being Orders Nos [ ] and [ ]).

Some arbitrators prefer to recite rather more detail, such as (if settled at that time) the date and other arrangements for the hearing, the sequence, extent of, or restrictions on the exchange of, statements of case, and so on. If anything of consequence stems from them, further details could be given of any of those procedural orders. If not clarified elsewhere, further detail could also be given as to the costs sanctions referred to in this illustration.

In cases of substance there might be several further meetings, sometimes principally procedural (akin to a basic preliminary meeting) or to monitor progress, sometimes principally to hear one or more contested oral applications for an arbitrator's Order. In cases of substance there can, if warranted, be a 'pre-hearing review' meeting, the main purpose of which is to ensure that all matters are proceeding so as to facilitate the effective conduct of the hearing. Once again, the fact of and/or the details or consequences of such a meeting should be recited *only* if and to the extent that they impinge upon or necessarily explain or clarify some aspect of the award.

5.3.4.9 Consequent and subsequent applications and orders [B2(i)]

This sub-section gives an opportunity to record, *to the extent relevant to the award*, arbitrator's Orders resulting from the meeting, or as a result of subsequent application.

In cases of substance or complexity, relevant matters could be set out here rather than being combined in the 'Meetings' item as above. In particular, any orders which were later the subject of peremptory orders (see B2(n) below) would be adequately outlined here.

Whilst in this illustration these matters are dealt with very briefly under B2(h) above, the type of matters likely to arise include (again to be recited only if and to the extent that they impinge upon or necessarily explain or clarify some aspect of the award):

(1)   Attended hearing or 'documents only'   [B2(i)1]

If the *manner* of receipt of evidence (by attended hearing or by submission of documents only) is not inherent from the outset (such as under any applicable rules) and recited earlier, those matters are likely to be settled at a preliminary meeting or its equivalent. That would commonly be recited here. See 9.3.4.5(1) for illustration.

(2)   Type of statements, timetable   [B2(i)2]

It is often sufficient simply to record the *nature* of such written statements as have been agreed or ordered to be exchanged. They might be formal pleadings akin to litigation, or statements of case, or in some other form. The fact of exchange of such documents can be dealt with briefly in a subsequent item (see B2(j)1 below), or might be combined, with this reference to them having been provided for. Some arbitrators prefer to set out a chronological list of the requirement and the fact of exchange, sometimes paraphrasing the essential content. In some circumstances that can be a suitable or convenient alternative format.

(3) Disclosure and production of documents [B2(i)3]
In some instances both parties have all the documents they need, so that no Order has been requested of the arbitrator, in which case no reference is needed in the award. In other instances a party is known, or alleged, or is thought possibly to have or to have had, documents which are likely to be necessary or helpful to the other party's case. If, following application, the arbitrator has issued one or more Orders in this respect, the extent to which that fact is recited at this point in the award depends substantially upon whether or not there were problems of non-compliance or the like (see for example B2(j)1, (k)4, and (m)).

In this illustration there had been only a limited order for disclosure of documents (e.g. one or more *specific* documents). The only subsequent problem was a failure to produce a document timeously, referred to briefly under B2(h) and (k).

(4) Exchange of witness proofs and experts' reports [B2(i)4]
(5) Meetings of experts [B2(i)5]
Once again, unless problems arose in connection with such exchange and/or meeting(s), the brief reference under B2(h) will often suffice. See 9.3.4.6(1) for illustration of a meeting of experts – and the result.

(6) Language and translation [B2(i)6]
The use of language other than English is fairly uncommon in arbitrations in the UK. That being so, if evidence *is* to be given, or documents produced, in a language other than English it would be prudent, whether or not subsequent problems arose, to recite that fact, together with the arrangements for translation and for dealing with any objections to the validity of translations. The need does not arise in this illustration.

(7) Questions to be put [B2(i)7]
This is not the book in which to discuss at length the ramifications of section 34(e), except to mention that it is phrased in wide terms: 'whether and if so what questions should be put to and answered by the respective parties and when and in what form this should be done'. It thus covers both oral and written questions and oral and written answers. It can include answers to be given by a written statement or by sworn affidavit. It can cover pre-hearing questions, and questions (examination) at the hearing. Written questions requiring written answers, in whichever form, are sometimes in arbitration still termed 'interrogatories'. (See 9.3.4.5 for illustration.)

At this stage of the award all that is necessary is concisely to set down what, if any, order(s) had been issued providing for such questions. Oral examination and cross-examination is so common-place that there is rarely need to make special mention of it unless problems have arisen. It will often be inherent in the matters set out at C(c) below.

(8)   Other relevant matters   [B2(i)8]
Some arbitrations can give rise to unique applications and/or Orders. This sub-heading is simply a prompt to pause and consider whether other matters should be recited in this award.

5.3.4.10   Resultant party action   [B2(j)]
This is where, again only to the extent necessary, matters arising from the Orders just referred to can be briefly set out:

(1)   Brief details of exchange   [B2(j)1]
The fact of pleadings or statements of case having been exchanged, of documents having being produced, of witness proofs and expert re-ports having been exchanged, and so on, is commonly referred to very briefly, unless problems in connection with those matters had arisen. In that case they, and any relevant consequence, would be mentioned to the extent that those matters impact upon the award. (See B2(j)3 below.)

Some arbitrators prefer briefly to list or tabulate the various stages of these submissions and exchanges, with or without brief precis.

(2)   Admissions or agreements, result of meeting(s) of experts [B2(j)2]
If as a result of inter-party discussion, or meeting(s) of experts, facts or figures have been admitted or agreed, that could conveniently be recited here. (See 9.3.4.6.)

(3)   Any resultant reduction in, or expansion of, the
        matters in issue   [B2(j)3]
Similarly, within the constraints of what had been referred to arbitra-tion (or, if such be the case, as has since been extended by written agreement of the parties), any reduction of the matters in issue (or any expansion thereof) could be recited here. If any issues had been re-solved, those issues, after recital here, would either be incorporated in Section H as a 'consent' element of the award, or (if that was, surpris-ingly, what the parties had requested in writing) would be excluded from the remainder of the award. In that last instance there should be a further recital stating that those specified issues had, by agreement, been removed from the arbitration.

In modest-scale cases, it can be convenient to merge these various matters (B2(j)1-3) in one item, perhaps, as here, where the nature of the claim concerning VAT had been amended (with no objection by the Respondent):

10 Statements of case were subsequently exchanged, as were witness statements, experts' reports and such documents as had been requested or ordered
to be produced. The consequence of delay in production is referred to at
paragraph 11 below. In the course of these exchanges and relating to Value
Added Tax, the Claimant amended that part of its claim to a request for a
declaratory award.

If the definition of the issue(s) stems directly from these interlocutory
matters (i.e. was or were not inherent in the initial submission) the
'Issue(s)' section shown earlier in these notes might well be better
positioned here rather than there.

5.3.4.11   Other directions and/or administrative actions   [B2(k)]
A number of matters (to some of which there might have been earlier
passing reference) can usefully be summarised here, shortly before
moving on to the hearing as such. These can include:

(1) Orders for security   [B2(k)1]
If one or more orders for security have been issued under section 38(3),
that should be referred to here or elsewhere, for such security (if given)
will have an impact in the operative section of the award (here Section
H). Any subsequent award of costs will need to have regard to any
security given and its form. See 9.3.4.7(1) and 9.3.11.3.

   If there has been failure to comply with such an order relating to
security for costs *and* with a subsequent peremptory order (see B2(n)
below) to the same effect, such that dismissal under section 41(6)
applies, that might happen at any time during the reference. In such
an instance the previously mentioned content of the award will obviously be abridged, depending upon what other matters, for instance a
counterclaim, remain. See 8.2.1 for dismissal following default under
section 41(6).

(2) Limitation of costs   [B2(k)2]
If there has been an order under section 65 directing that recoverable
costs, or any part thereof, shall be limited to a specified amount, the
fact of such order should be referred to here or elsewhere, as background to or explanation of the consequential limiting proviso which
should appear in Section H after consideration of its effect in Section G.
See those sections and 9.3.4.7(2) for expansion and illustration.

(3) Any valid 'costs in any event' agreement   [B2(k)3]
The parties might have agreed that one of them is to pay the whole
or part of the costs of the arbitration regardless of the outcome ('in
any event'). By section 60, that is only valid if made after the dispute
arose.

Imagine that a party seeks to rely on such agreement but the other party contends that it was made prior to the dispute arising and was therefore invalid. The arbitrator, failing post-dispute agreement between the parties on that matter, would need to consider submissions on the conflicting contentions and decide whether or not any such agreement was valid. Those background facts, and his determination, should be recited here or when dealing with costs (Section G). If, by contrast, there is an uncontested agreement in this respect, the arbitrator could recite that here, as background to his subsequent award on costs, which would give effect to that agreement.

(4)   Any 'costs in any event' directions   [B2(k)4]
Illustrative award paragraph 9 above makes reference to two orders 'with a sanction as to costs' . Assuming that those two orders provided that 'The Respondent shall pay the costs of and arising from this order in any event' or words to that effect, the entry here might be:

> 11   My Orders Nos [*number*] (failure to produce documents timeously) and [*number*] (abortive application) provided that the Respondent be liable for the costs of and resulting from those orders in any event.

The consequence of those orders would be dealt with in the operative section of the award (but see in Section G later).

5.3.4.12   Provisional orders   [B2(l)]
Provided that the parties had granted that power under the provisions of section 39, if at some date prior to this award the arbitrator had ordered relief on a provisional basis by way of payment (or disposition of property, or other available relief) relating to one or more issues which have not yet been finally decided, or had ordered an interim payment on account of costs, that should be recorded as a fact. The position of this item in the sequence of the award will, like others, depend upon the surrounding circumstances and other matters recorded. For instance, this particular point can often most suitably be recorded earlier with one of the items dealing with interlocutory matters. The final disposition of any such payment (or declaration, or other decision) must be determined by this or a subsequent award. This provision is inapplicable in this present case, but see 7.10.1 and 9.3.4.8.

5.3.4.13   Party default (and consequence)   [B2(m)]
Some defaults by a party might have been minor, had been dealt with at the time, and had no adverse effect upon the other party or the proceedings. These need not be recited unless *cumulatively* they have, or are likely to have, such effect that they could affect the arbitrator's decision on some matter to be dealt with in this or a future award in

the arbitration (including costs). Where they have, or are likely to have, such effect, they should be recited.

Other defaults can have adverse effect from the outset. A not un-common example could be a party's failure to comply with an order or direction of the arbitrator, say failure to allow inspection of a specified document. That default, and any consequences, would be recited. See below at B2(n) regarding resultant peremptory orders and see 9.3.5.1 for illustration.

Default can also arise during, or at the time for, the hearing (see B3(f) below). Whether it is recited here or under that later item will depend upon the nature of the default.

(1)   Inordinate or inexcusable delay   [B2(m)1]
In the event that there is such delay on the part of a claimant (or counterclaimant), provided that it has or is likely to have one of the consequences set out in section 41(3), the arbitrator may 'make an award dismissing the claim'. In that instance the comments at the second paragraph under B2(k)1 above, regarding abridging of recitals, will apply – and also see 8.2.1.

(2)   Failure to comply with any order or direction   [B2(m)2]
Subject to the previously-mentioned proviso that such matters should usually be recited only if they impact on or affect the understanding of the award, failure to comply with an Order of the arbitrator which then led to a peremptory order would be recited, either as such or combined with the next item (B2(n)) dealing with peremptory orders.

5.3.4.14   Peremptory orders   [B2(n)]
If, following appropriate default by a party, the arbitrator had issued a peremptory order under section 41(5), that fact should be recited. That is particularly so if the original default is likely to influence the arbitrator's decision on liability for costs. If there has been failure to comply with a peremptory order itself, thereby leading the arbitrator to apply any of the sanctions listed in section 41(6) or (7), that too should be recited. The situation does not arise here, but see 8.2.1 and 9.3.5.1.

5.3.4.15   Other relevant matters   [B2(o)]
This matter is mentioned again because, unless agreed otherwise between the parties and the arbitrator, the parties can agree *at any time* in respect of matters regarding which the 1996 Act gives that facility of agreement. That situation can, and does on occasion, arise immediately before or during the hearing. Such agreement might relate to the arbitrator's powers, or to procedural matters, in respect of one or more issues. Examples could be the reinstatement of an initially excluded power to take the initiative in ascertaining the facts

and the law concerning all or some issues; or about procedures, such as agreeing that the strict rules of evidence should not apply in respect of certain issues.

The arbitrator should record such agreement(s), either here (if there was good reason to recite it at this point) or linked with the earlier items on such matters (which would often be the more logical).

An example of matters recorded under this item might be:

> 12 The parties agreed that in view of the relatively modest amount in dispute there was no need for a pre-hearing review, but I was notified by a joint letter dated 20th May [year] that they had agreed to limit the length of the Hearing to one-and-a-half six-hour days plus a site inspection; the time to be shared equally between them, but 'with a reasonable amount of give and take and with any differences as to the sharing of time being resolved by the arbitrator on an instant but fair basis'.

Section B3

5.3.5 *Hearing*   [B3]

Again, a sub-heading:

> **Hearing**

This is another example of a situation where, depending upon the circumstances, two or more of the items in this section of the checklist can be conveniently combined. That depends to some extent on, for instance, whether the matter is dealt with by an attended hearing, or on submitted documents alone, or a hybrid procedure. It is likely that in many situations it will be convenient to merge sections B3 and C, but they are kept separate in the checklist and in this illustration solely to keep 'administrative' matters separate from consideration of the dispute as such. For this illustration it is assumed that there was an attended hearing, and again for present purposes the various items are separately illustrated.

5.3.5.1 Transfers   [B3(a)]
Again, this checklist sub-heading is simply a reminder to adjust the checklist to suit the particular award.

5.3.5.2 Hearing   [B3(b)]
Whilst it is not usually necessary to recite just where and when the hearing took place, such recital can help put matters into perspective, particularly if there were problems by way of delays, non-attendance, or the like. An appropriate entry might be:

> 13   The hearing was held on Tuesday and Wednesday 24th and 25th July [*year*] in the Chamber of Commerce Council Chamber at Weiringhampton.

5.3.5.3   Failure to attend/ex-parte proceedings   [B3(c)]
In the event that within the provisions of section 41(4) the arbitrator had proceeded ex parte (or, where matters are to be dealt with in writing, has proceeded without written evidence or submissions from the defaulting party), that should be stated, briefly giving the reasons and background. (See 9.3.5.1 for illustration.)

5.3.5.4   Representation and witnesses   [B3(d)]
Unless to be recited later when dealing with submissions and evidence in Section C, there could now be a brief indication of who represented the parties and who appeared as witnesses, giving their designations; and of any expert witnesses, indicating the area of expertise in each case. If, as is common, witness statements (and/or experts' reports) were taken as read (i.e. not repeated orally except, if allowed, by way of confirmation and possibly emphasis, explanation or expansion) as evidence-in-chief, that would be stated, mentioning any facility for oral confirmation or expansion-in-chief and whether opening and closing addresses were oral or in writing. This might be:

> 14   Both parties were represented by Counsel, Ms Susan Wellbie for the Claimant and Mr Justin Blackwelle for the Respondent. Opening addresses had been submitted in advance by both Counsel, and closing submissions were also submitted in writing in sequence.
>
> The various witnesses were:
>
> Called by the Claimant:
>
>   Witnesses of fact:
>
>   Giles Jay Wyttle   : Technical Director
>   George Amble   : Multi-skilled technician
>
>   Expert witness(es):
>
>   Arthur Worsnip   : Hi-tech set specialist
>
> and by the Respondent:
>
> Witnesses of fact:
>
>   Hyram Trimble   : Managing Director
>   James Bull   : Stage Manager
>
> Expert witness(es):
>
>   Jon Chekov   : Independent producer
>
> Witness statements and experts' reports having been exchanged, they were, as had been agreed, taken as read as evidence-in-chief, but with oral confirmation and limited expansion.

If, as here, counsel appeared, they would in a substantial case often be accompanied by junior counsel. It would commonly be the case that the instructing solicitors would also be present. The extent to which their respective presence should be recited depends substantially upon whether they took a 'speaking' role. If to be named at all, non-participants can be listed as 'also present were...', but that would be little more than a courtesy mention and without relevance unless to explain some other matter which arose.

The reference to witness statements could alternatively be recited in Section C.

If not inherent in references to the expert witnesses in Sections C or D, their status and qualification could be set out here or in the list just illustrated.

5.3.5.5   Oral evidence   [B3(e)]

Oral evidence is commonly taken on oath, or on affirmation if a witness reasonably objects to swearing an oath. An example might be:

> 15   Oral evidence was generally taken on oath. Instances where a witness requested instead to affirm are mentioned in the relevant section of this award only if and where necessary to explain my reasoning or conclusion.

5.3.5.6   Default during hearing   [B3(f)]

In just the same way as there can be default by a party during the preparation for a hearing, this can occur during a hearing. Even at that stage there can be such matters as failure to produce documents, or disruptive applications, or late appearance, and so on (and see B2(m) above). Those situations, and their consequences, should be recited in the same way as other relevant defaults. Such matters can be set out here, or alternatively in more or less chronological sequence when dealing with submissions and evidence later.

5.3.5.7   Inspection   [B3(g)]

If the arbitrator had also inspected the subject matter of the dispute, or the location, or other relevant things, that should be recited, stating whether alone, or if accompanied, by whom. At this stage the reference to inspection would commonly be little more than a statement that it had occurred, but it would be in order, if it would not disrupt the flow of purely factual background, to state what had been seen, i.e. without any analysis or reasoning. Analysis and reasons come later. In this case the item might be, say:

> 16 On completion of the hearing on 25th July [*year*] I visited Theatre STL, accompanied by Mr Worsnip and Mr Chekov, and inspected the work of decoration in question and the sets. My attention was drawn to:
>
> (a) the nature of the mobile sets and their integral finishes;
>
> (b) the extent and degree of complexity of the decoration to all spaces behind the line of the proscenium opening (including stage equipment and gantries); and
>
> (c) the nature of access to those areas;
>
> the purpose of my inspection being solely to familiarise myself with these matters, not to make decisions on the basis of that inspection.

5.3.5.8 Agreed list of issues [B3(h)]

There can be further clarification of, or amendment to, or reduction in, the previously-defined issues, or notification of payment made during the reference. That would preferably be recorded by adaptation of the earlier item (at B1(h) or B2(g)), suitably annotated and possibly with a brief cross-reference here.

5.3.5.9 Other matters [B3(i)]

This item is simply to generate a pause for thought. Is it either essential, or helpful to the understanding of the award, to set down any further background? If so, but only to the extent that it is so, that should be done.

Section C

5.3.6 *Submissions and evidence* [C]

It was indicated in a fore-note to Section C in the checklist that where the arbitration is conducted on a 'documents-only' basis, and in some other circumstances, it can be convenient to deal with submissions and evidence prior to proceeding with the analysis. For attended hearings, however, the submissions and evidence can conveniently be combined with the analysis etc. in Section D. Many arbitrators (including the present author) prefer to use that combined sequence when related to attended hearings unless other factors indicate differently. That combination of C and D is adopted here, using the checklist as simply that, not as an indication of sequence.

First the heading, if a separate section had been adopted:

Submissions and evidence

Whichever sequence option is adopted, it is useful to both the drafter and subsequent readers to locate a summary of contentions in fairly close proximity to the analysis and decision-making. For present purposes, it is dealt with now, but in effect as part of a combination of Sections C and D.

As in some instances Sections C and D might be kept separate, each sub-heading in the checklist is briefly commented on here whether or not illustrated.

5.3.6.1 The contentions [C(a)]

The contentions of the parties on the various issues of claim and counterclaim can usually be copied from, or paraphrased from, the parties' 'pleadings' (not to be confused with their advocates' later submissions). They can be set down here in a variety of ways:

(a) As the *fact of them making* such contentions is not itself contentious, they can conveniently be recited in one of the 'background' sections (e.g. B1(h)), issue by issue (or sub-issue by sub-issue); *or*

(b) They can be incorporated in, or form a preface to, this 'Submissions and evidence' section; *or*

(c) Under the heading of each issue in turn, the award can deal in sequence with contentions, submissions and evidence, the arbitrator's analysis, consideration and decision.

The option chosen depends partly upon the circumstances of the case and partly on the individual arbitrator's preference. Any one arbitrator might well use different options in different circumstances.

Procedure (b) is adopted here, but solely to better identify the various separate functions.

In cases of any substantial complexity, issues and sub-issues will usually have been scheduled in one form or another, whether by simple listing, by Scott Schedule[1], or by other means. Any such detailed scheduling should be identified by reference here (for cross-referencing and detailed consideration in the 'Analysis and decisions' section later).

Whilst in this example contentions are taken issue by issue, for convenience sub-issues (a) and (b) are taken together, being closely related:

---

[1] A 'Scott Schedule' is a schedule which, item by item, sets out (usually in horizontal juxtaposition) the contentions of the respective parties, with a column for the arbitrator's use. Such schedules usually, but not necessarily, relate to matters of account, or quantity, or the like.

17 **Relating to sub-issues (a) and (b):**

(Was the disputed work included in the Respondent's letter of enquiry dated 1st October [*year*]? If not, was it specifically included in the Claimant's quotation dated 10th October [*year*]?):

The Claimant contends that:

(a) the backstage redecoration, not forming part of the sets as such, was not referred to in the letter of enquiry upon which the Claimant's tender was based, nor was it included in that tender; further, that such exclusion was apparent on the face of the tender;

(b) the Respondent was at all relevant times aware that the said works of decoration constituted work of a nature not normally executed by the Claimant;

(c) whilst pre-invitation there were wide-ranging discussions between the parties concerning the whole of the stage area and its equipment, only the mobile sets were mentioned in the letter of enquiry;

(d) the said works of redecoration were not a necessity for the proper functioning of the mobile sets.

and so on, briefly setting out the claimant's contentions, then:

The Respondent contends that:

(a) it was at all times clearly understood and acknowledged by the Claimant that the Respondent required the backstage areas to be decorated at the same time as the installation of the sets, and there was never any question of that work being done by others;

(b) there was no specific exclusion of that work in the Claimant's quotation dated 10th October [*year*];

(c) there was a necessary inference to be drawn that the said works of redecoration were an essential prerequisite for the functioning of the mobile sets.

followed by the remaining issues and related contentions (again here combined):

18 **Relating to sub-issues (c), (d) and (e):**

(If neither, did the Respondent subsequently order that work? If so, did the said work constitute either extra work or a collateral or other contract entitling the Claimant to extra payment? If so, what amount was a fair charge for that work?)

The Claimant contends that:

(a) the Respondent's agent, Mr Hyram Trimble, who had signed the letter of acceptance of 5th November [*year*], gave clear oral instructions for the execution of that work and when so doing was aware that the Claimant's

schedule of work included in the accepted quotation made no reference
to that work;

(b) it is entitled to extra payment for the said work;

(c) the amount claimed represents a fair quantum meruit evaluation.

The Respondent contends that:

(a) whilst acknowledging that Mr Trimble was its accredited agent, his
comments to Mr Amble had constituted nothing more than clarification,
not intended to be an order for extra work;

(b) in consequence no extra payment is due therefor;

(c) if contention (b) is rejected, a fair figure is little over £10,000.

See 6.4 for counterclaim illustrations.

5.3.6.2   Opening and closing submissions   [C(b)]
The parties' contentions are usually supported by submissions of law
(or such other basis, if any, as is relevant under section 46(1)(b)) and by
oral submissions by Counsel or other advocates. If the award is to have
a separate 'Submissions and evidence' section, then to the extent that it
is necessary to summarise those submissions (but only to that extent,
i.e. without at that stage considering them), that can be done here.
Otherwise these matters can be incorporated in the later 'Analysis and
decisions' section, as is done in this illustration.

5.3.6.3   Witnesses and evidence, if not covered in Section B: [C(c)]
(1)   Identification of witnesses   [C(c)1]
If not done in Section B3 (as was done in this illustration) this could be
dealt with here. If the arbitration was conducted on a 'documents only'
basis, and if such documents included witness statements, those witnesses might well not have been identified in Section B3. If not dealt
with there, that too can conveniently be done here. If such statements
were by way of sworn affidavits, that should be stated.

(2)   Précis of relevant evidence   [C(c)2]
If there is a hearing, a précis of evidence can conveniently, as here, be
incorporated under Section D. The précis should be restricted to that
which is essential to the comprehension of the award.

   If on a 'documents only' basis, it can be helpful to include a short
précis of the relevant evidence here, by way of background to what
follows in Section D. Alternatively, if desired or more convenient in
the particular circumstances, that too can be incorporated in Section D.
In a documents-only situation of substance the equivalent of 'admissions on cross-examination', is admissions resulting from answers to

'interrogatories' (i.e. written questions posed to the witness – see at B2(i)). For brief illustration regarding relaxation of strict rules of evidence, see 9.3.6.1.

(3) Equivalent description in ex-parte proceedings [C(c)3]
Where it has been necessary for the arbitrator to proceed in the absence of a party or its representative under the provisions of section 41(4), the manner in which that has been done should be adequately set out.

5.3.6.4 Inspection – if not covered in Section B [C(d)]
(1) Purpose of inspection [C(d)1]
Having, in this illustration, already dealt with a basic inspection (at award paragraph 16, under B3(g)), this sub-heading is shown again here simply because in some small arbitrations, including some 'consumer schemes', the procedure is one of submission of documents plus an arbitrator's inspection. In such schemes (see 8.3.2) the inspection can (if so designated in the scheme) take on a different and more substantial significance. In those circumstances, the thing in issue is, in effect, proffered in evidence (see C(d)2 below and brief comment at 9.3.4.5).

(2) Result of inspection [C(d)2]
If the purpose of the inspection was to adduce evidence (the thing inspected) and to make a decision on that basis (e.g. in some 'quality' disputes), that would be so described – with the finding and decision being shown in Section D.

5.3.6.5 Other matters [C(e)]
Again, this item is simply to generate a pause for thought. Is it either essential, or helpful to the understanding of the award, to set down any further matter pertaining to submissions and evidence? If so, but only to the extent that it is so, that should be done.

Section D

5.3.7 Analysis/findings/reasons/decisions on the substantive issues [D]

Too many award-drafters, whether 'new' arbitrators or examination candidates, write the foregoing background, and what follows, at excessive length, with unnecessary blow-by-blow accounts of meetings and verbatim repetition of what the witnesses said – but all too little as to *how and why* the arbitrator has reached a conclusion. However, if (as is standard under the 1996 Act unless agreed otherwise) a reasoned award is required, this present section, or a separate 'reasons' section, *must* clearly demonstrate those reasons. It is inappropriate to set out all the reason*ing*.

The arbitrator must determine each and every issue referred to him and such determination must be apparent and unequivocal.

5.3.7.1   List of issues   [D(a)]
In a lengthy award, it can be helpful to a reader if the description of the issues is repeated here, unless they have been recited in such close proximity as to make such repetition unnecessary. Alternatively, this section could begin with a brief cross-reference to the paragraph where they were listed. In this illustration (see paragraph 8, at B2(g) above) there is only one issue, made up of five sub-issues. These sub-issues are dealt with in turn, under appropriate headings. (See illustration following D(c)6.)

5.3.7.2   Common ground   [D(b)]
In some instances items of common ground (i.e. matters of fact or law on which the parties are in agreement) span across or influence all or most issues. In such a case they can to advantage be dealt with together under a suitable heading before moving to the analysis as such. In other instances the items of common ground relate to single issues. In that case it is likely to be most convenient that they are set out within the consideration of the issue to which they relate. (See award paragraph 20.)

5.3.7.3   Analysis of submissions and evidence   [D(c)]
The award illustration starting at paragraph 19 below is shown after the remaining discussion of the whole of checklist items D(c)(1)–(6) inclusive.

(1)–(3);(5) Facts/law/application/decision   [D(c)1–3;5]
It is convenient here to consider these sub-sections together, before the remainder. The reasons for an arbitrator's decision might be based solely upon one or more facts, or solely upon one or more conclusions on the relevant law, its interpretation and application, or, much more commonly, upon an interplay between the two, the application of one to the other. The need of the parties to have reasons is not synonymous with that of the court. The court needs the arbitrator's reasons only in the event of application or appeal being made to it. By contrast, the parties need to understand (sometimes for future commercial purposes) *why* and *how* a decision has been reached – on fact as well as law. Thus a reasoned award needs to set out, logically and issue by issue:

(a)   in respect of relevant facts, the reasons for finding those facts (i.e. for determining alleged facts to be correct, or for rejecting them);

(b)   in respect of law, the reasons why a particular interpretation has been adopted; *and*

(c)   in respect of the overall decision on each issue (on liability and on quantum), the reason (if not self-evident from the decisions on fact and/or law) why the application of that law to those facts gives that determination.

The verb 'find' is used by some arbitrators to indicate facts 'found' – i.e. determined to be valid – but is also sometimes used in the context of the final decision – 'I find for the Claimant'. There should be no confusion when each is considered in its context.

As mentioned earlier, whilst some arbitrators set out *all* background at the beginning of the award, others might incorporate some of those matters in their consideration of the individual issues to which they particularly relate. Similarly, if several issues depend, in part or whole, upon common facts or upon common matters of law, it can be convenient to deal with those matters of fact and law first, so that the conclusions can then be applied to the several issues. Where each issue depends upon matters of fact and law which are not common to other issues, the determination of those matters can often be most conveniently dealt with within each particular issue, whether under separate sub-headings of 'fact' and 'law', or combined. That last approach is adopted in this illustration, merging several parts of the checklist and not seeking to segregate determination of all matters of fact from those of law.

That is not to suggest that the arbitrator should do anything other than state 'I find as a fact that...' (or words to that effect to suit the syntax) where there are clearly defined matters of fact as such. Findings of that kind can often, however, be incorporated in the text without interrupting the natural flow of reasoning by artificial segregation.

It is also necessary to bear in mind (unless the parties have excluded the right to appeal on a point of law) that section 69(3)(c), dealing with restraints on leave to appeal, starts: 'that, on the basis of the findings of fact in the award...'. It is necessary, therefore, by one means or another, to make it quite clear what facts have been 'found'.

In this illustration, when reviewing evidence, instances are intentionally shown of short *verbatim* extracts of a witness's oral evidence (e.g. at paragraph 29). It is important to recognise that such quotations can only be included when the arbitrator has a word-perfect note of precisely what the witness said. Otherwise it is safer to accurately paraphrase that part of the evidence, for an incorrect quotation could provoke a challenge to the award. On occasion, however, a verbatim quotation can be more telling.

It is sometimes the case that counsel (or other advocates) notify the arbitrator during the hearing that they are agreed that there is no need to address one or more legal issues. This can happen, for instance, where there are 'in the alternative' pleas and one or more of the alternatives are dropped. An example is shown in this illustration at paragraph 32, and of admissions at paragraph 37.

The decision on each issue (and sub-issue) should be clearly shown, and it can be useful (and particularly so if an award is more complex than this illustration) to show the various decisions on sub-issues and issues in bold type, thus 'punctuating' the award.

For illustration of determination on the basis of 'other considerations' (section 46(1)(b)), see 9.3.7.1. For brief consideration of appended reasons, see 9.3.7.2.

(4)   Interest as special damages   [D(c)4]
Part (4) does not arise here, but constitutes a reminder in the checklist. (See Chapter 4 at 4.3.3.1.)

(5)   Decision on substantive issue(s)   [D(c)5]
In the illustration below, this item is linked with D(c)(1)–(3). See above and illustration below.

(6)   Counterclaim or set-off   [D(c)6]
If there were a counterclaim, this matter of set-off or not would be considered here, i.e. before dealing with the remaining matters. See Chapter 6, at 6.4, for illustrations of counterclaim/set-off.

Illustrations relating to D(a)–(c) above
Reverting to the illustration and here adopting the sequence of developing the reasons in the body of the award (but also see 3.1.4 and 9.3.7.2), first there would be a suitable heading, which might be:

My consideration of the issues

followed by consideration of the issues in turn, but here a single, five-part issue:

The sole issue: *Is the Claimant entitled to additional payment for executing the work in dispute and if so how much?*

followed in turn by the sub-issues:

> **Sub-issues (a) and (b): Was the disputed work included in the Respondent's letter of enquiry dated 1st October [year]? If not, was it specifically included in the Claimant's quotation dated 10th October [year]?**

Whilst not essential, there can here be a reference to some sub-issues being taken together:

> 19   It is convenient to take these two sub-issues together.

If there is only a short list of matters of common ground, that entry can simply follow in sequence. If there is a lengthy list of such matters, it is preferable to show them under a suitable sub-heading, either interjected here, or as a free-standing section preceding this 'analysis' section. If the first-mentioned sequence is adopted, it might be as simple as:

> 20   It is common ground that:
>
> (a)   the decoration to backstage surfaces was executed by the Claimant;
>
> (b)   a colour schedule was passed to the Claimant by the Respondent;
>
> and it is not in issue that:
>
> (c)   the letter of enquiry as such made no specific mention of decoration to backstage surfaces.

That might continue with:

> 21   The Respondent's Managing Director, Mr Trimble, contended that he had repeatedly during earlier discussions mentioned the required visual impact between the mobile sets and the blank rear wall. He was vague as to just when those discussions took place.
>
> 22   To the extent that, if at all, such discussions were relevant, I accept as the more convincing Mr Wyttle's evidence (and I find as facts) that those discussions were not only of a general nature and preceded any invitation to submit a quotation, but also related to the visual link between the proscenium opening, the curtains (which were never any part of the work to be quoted for by the Claimant), the sets in their various configurations, and the rear (and any other visible) walls.
>
> 23   Mr Blackwelle, for the Respondent, submitted that the contract was partly oral ('a meeting of minds in those discussions') and partly written ('the enquiry, the quotation – i.e. the offer – and the acceptance'). Ms Wellbie, for the Claimant, argued that the parties by those written documents had put their agreement into precise terms.

24 The letter of enquiry stated 'Following our various meetings, I am setting out the parameters to our requirements: a dimensioned sketch design to be developed by you; functions; rotates and other movables required; minimum needs re power outlets; computerised facility requirements; and colours of exposed – i.e. visible in any of the configurations – parts.'

25 The quotation incorporated that letter by reference, and added technical specifications and outline working drawings; it made no reference to back-stage redecoration, nor to discussions nor to other agreements. Those matters are not in contention.

26 It is clear that the enquiry and quotation did put the agreement in precise terms. The enquiry was in sufficient detail as to performance and extent as to make it inconceivable that other work of any consequence could have been intended but not mentioned. The works of redecoration (much of which were at all times, even on an open stage, not visible from the auditorium) were clearly not an essential prerequisite for the functioning of the mobile sets, there was no such inference to be drawn, and I so hold. It is therefore inappropriate for me to consider oral evidence of some alleged different or additional intent, but even if that were not so, I am satisfied from Mr Wyttle's evidence that there was no such intent by either party.

27 In consequence, I hold that the letter of enquiry must stand by itself and be so interpreted.

28 ***I determine that the disputed work was not included in the said letter of enquiry, nor was it included in the Claimant's quotation, whether specifically, by reference, or by necessary inference***.

Sub-issue (c): If neither, did the Respondent subsequently order that work?

29 Mr Amble, the Claimant's technician who supervised the site assembly of the sets, gave evidence that on many occasions Mr Trimble said things like *'I want you to paint these walls and those others before you finish assembly of the sets'*, to which he had replied to the effect of *'I know nothing of that'* in response to which Mr Bull, the Stage Manager, had been asked by Mr Trimble, in his presence, to give him (Mr Amble) a colour schedule. That colour schedule (document C27) is headed *'Paint the surfaces to the following colours:'* and was signed by Mr Trimble. I found Mr Amble's evidence convincing, but that of Mr Trimble and Mr Bull vague and evasive on this point. Mr Amble said that he took the repeated requests by Mr Trimble and the reaction to his own response, including the issue of that schedule, as constituting an order to do the work. He also told us that he had then said *'OK. I'll ask the boss to raise a works number'*, to which, he says, he got the response *'Yes, do that'*. Once again I find Mr Amble's evidence convincing.

30 In consequence, I hold that Mr Trimble's words and actions led Mr Amble, quite properly, to believe that extra work had been ordered. Neither Mr Trimble nor Mr Bull did anything to disillusion him in that belief. In those

circumstances any reasonable person would justifiably have taken it that he had received an oral order.

31 ***I determine that the Respondent did subsequently order that work.***

Sub-issue (d): If so, did the said work constitute either extra work or a collateral or other contract?

32 It was made clear to me by both Counsel that the term *'extra work'* was intended simply to refer to work ordered as a variation of or under the contract as contrasted with a separate contract of some kind. They agreed that nothing now hangs on the distinction, and that, if I find the work to be extra work, I am not requested to determine its precise contractual status.

33 I have no hesitation in concluding that the work was 'extra' to the extent that, as I have found, it was additional to the contract work, and it was ordered by the Respondent – and that the Claimant is entitled to extra payment for it.

34 ***I determine that the said work did constitute extra work of a nature in respect of which the Claimant is entitled to be paid.***

Sub-issue (e): If so, what amount was a fair charge?

35 Page D43 of the agreed bundle was a copy of part of the Claimant's computation of the amount charged and claimed. I pointed out that there appeared to be a duplication of the entry for one operative and it was conceded by Ms Wellbie that that was so. That would have had the effect of reducing the amount claimed to £18,150.00.

36 Mr Worsnip, the expert called by the Claimant, had considerable experience not just of 'mobile' set design but of traditional sets and of re-equipment and (of specific relevance here) of refurbishment of backstage areas. He had analysed the successful bids relating to such work at six theatres with which he had been concerned and had broken down the various figures in a manner which gave broad unit prices and facilitated comparison with the redecoration at Theatre STL. On that basis he gave as his opinion that the work here in dispute, had it been the subject of competitive tenders, would have cost approximately £16,000. His assessment of the impact of the difficulty of access caused by the presence of the Claimant's set erectors simultaneously executing their contract work amounted to a further £750.00. He said that if works had perforce to be executed by non-painters it could be expected to be more costly, perhaps of the order of 5% to 7½%. On the other hand, Mr Chekov, the expert called by the Respondent, simply contended that he could have had the work done for no more than £10,000, but produced no justification for his contentions. In cross-examination, he admitted that he had little experience of this kind of work.

37 Ms Wellbie asked me to have regard to the compounding effect of the various unusual factors: an unexpected order for this work, the need to complete by the originally-intended date, attempts to find and the non-availability of local painters, the remote nature of the location, and the enforced use of technicians who were not general painters. She contended that this compounding factor

fully accounted for the difference between Mr Worsnip's figures and the amount claimed. Mr Blackwelle confirmed that the Respondent admitted the points concerning the non-availability of local painters and that the work had of necessity to be executed by the Claimant's technicians, but did not accept that the Claimant's charge was fair. I find that there is weight in Ms Wellbie's contention, even if not fully to the extent suggested. Having regard to these various matters, I have determined that a proper amount for this work, excluding any applicable VAT, is £17,100.00.

38 *I determine that a fair charge for the work in question is £17,100.00.*

It is then convenient and appropriate to pull the sub-issues together as a decision on that full issue. There is no mandatory or standard form of words other than that they should be unequivocal and lead logically to the operative part of the award. Where, as in this present instance, there is but one issue, the award might move direct to the operative part without an intermediate decision on the issue as such. Otherwise, there could here follow (where, if there were more than one issue, the issue number would be stated):

My decision on this issue.

39 **I determine that the Claimant is entitled to additional payment, in the sum of £17,100.00.**

5.3.7.4   Summary of decisions on substantive issues   [D(d)]

If there were more than one issue, the award would continue issue by issue, dealing with the facts and the law, determining liability (or not), then if appropriate deciding quantum (or appropriate remedy) relating to each.

The various decisions can now be brought together and summarised. This is in readiness for the next task of deciding whether or not interest is applicable on such amounts as have been found to be due, and then on how costs are to be awarded. In other than relatively modest cases, however, interest (possibly, if complex) and costs (or at least the quantum of costs) are likely to be reserved to later submissions and a further award.

In this single-issue award, no separate summary as such is needed.

In some instances where there are a number of monetary claims, interest (when considered – see Section F) might be determined to run for different periods of time, or at different rates, on the various amounts awarded. If that is to be the case, any summary at this point can be no more than a list, i.e. untotalled. Such a list can nonetheless be a helpful 'pulling together' in a complex or lengthy award. In those circumstances, or where there is a counterclaim (whether or not it constitutes a set-off), any financial summary can only be completed when interest computations are dealt with, i.e. later in the award.

For a variety of illustrations, see 6.3 and 6.4.

Had there been several decisions, followed by one or more issues which were to be set off, the former could here be listed and totalled, then the set-off items similarly collated and deducted from that total – all in the one summary. Sometimes that deduction can be done before considering interest, i.e. if any interest will simply apply to the net result of claim less set-off. Sometimes if, as in the illustration at 6.4.1, the period and rate of interest on the claim and set-off differ, any interest must be applied to the claim and set-off amounts awarded separately, before arriving at a net figure.

5.3.7.5   Amounts paid during period of arbitration   [D(e)]
Payment (if any) of part of the amount claimed in and outstanding at the commencement of the arbitration might have been made at any time. There might have been a provisional order under section 39, or the respondent might have voluntarily made one or more payments. The fact of such payment(s) might have been recited earlier; if not, that could be done here – together with a further entry in the summary, so as to arrive at a net figure for the substantive award (prior to any addition of interest – but note the comments under 5.3.7.4 above). See 7.9 for further illustration.

Section E

5.3.8   *Value added tax implications*   [E]

5.3.8.1   Dealing with value added tax   [E(a)-(c)]
A variety of VAT-related matters can arise in connection with an award. These are discussed in Chapter 4 Section E. In this present illustration the claimant has requested a declaratory award on that element. That method of dealing with possible VAT could be likely, in any event, if there was doubt in any particular arbitration as to whether any or all of the work in question was a taxable supply and whether part or all was standard-rated for VAT. See 9.3.8.1 and .2 for further illustration of other VAT situations.

Although this 'declaration' (in fact a kind of unquantified monetary award) will appear in Section H, i.e. the 'Operative Section', it is still necessary to show here *why* it is awarded. A brief entry here (with a sub-heading, unless it follows immediately after the relevant substantive decision or a summary of decisions) might in the present circumstances be something like the following, here assuming that at the time of the amendment of its statement of claim (award paragraph 10) the claimant had given this as its reason (i.e. the negotiations mentioned below):

Value Added Tax

40 There was no dispute between the parties that if I were to award any sum to the Claimant, Value Added Tax would then apply to some part, or all, of the amount awarded. {However, as negotiations were still in hand with HM Customs and Excise as to whether elements of the work (being to a listed building) were in fact subject to that tax, it would be impossible at present to ascertain the amount of the tax.} I have found a sum to be due under the substantive claim and in consequence hold that the Respondent is in addition liable to pay such amount in respect of VAT as is in due course found to be properly payable under the VAT Regulations, together with *interest if such payment is delayed.

Section F

5.3.9   *Interest*   [F]

5.3.9.1   **Interest within the award**   [F(a)1,2] [F(b)]

Factors affecting the award, or not, of interest are discussed in Chapter 4 Section F. The enumerated notes below refer to the 'tasks' in that section, using the same references ((i)–(xi)), rather than to the checklist. (Tasks (xii) and (xiii), i.e. post-award interest, are considered at 5.3.9.2 below. See 6.4 for illustrations relating to interest where there is a counterclaim, and 7.9 for interest under section 49(3)(b) on an amount claimed in etc. and paid during the arbitration.

Unless to be reserved to a later award, the decision on the substantive issue(s) should be followed by consideration of whether or not a successful party, including a successful counterclaimant, is entitled to have interest added and, if so, computation of that interest. In common with the substantive issues, it is necessary to give reasons for the award, or not, of interest – and to specify the basis. Brief reference is made at 4.3.3.1(4) regarding differing views as to such interest on an award of costs.

As mentioned at 4.3.3.1(4) and at 5.3.7.4 above, it is possible for interest on different issues to run for different periods and/or at different rates. That can clearly affect the question of at what stage and in what manner the amounts (if any) awarded are summarised, and how interest on the various parts is set down. When carrying out those tasks, it is essential to keep in mind the manner in which the operative section H will be set out. In a particularly complex situation, it can be helpful to compute the various interest figures, then total the

---

* See note at 5.3.11.6.

substantive amounts awarded and similarly total the various interest figures. The award in section H could then give those two figures, but cross-referenced to this point in the award, e.g. 'as computed at paragraph [*number*]'.

In this present illustration and regarding 'tasks' (i)–(xi), (but these matters only need to be recited if and to the extent that they affect the decision):

(i) interest was claimed under section 49 (only);

(ii) there were no constraints upon the arbitrator's powers to award interest;

(iii) there were no complexities, nor matters requiring further party submissions, to warrant interest being reserved to a future award;

(iv) the claimant had been held out of its money, and the award of interest is appropriate (see also (viii) below);

(v) the matter of set-off does not arise;

(vi) there had been no payment relating to amounts claimed in and outstanding at the commencement of the arbitration;

(vii) the matter of non-discretionary interest does not apply;

(viii) the date upon which payment should have been made is clear from the contract terms (and see (iv) above);

(ix) for this illustration, and whilst parties will often be far from so co-operative, it is assumed that:

    (a) the claimant's banking situation varied approximately equally between deposits and borrowing;

    (b) at the hearing the claimant had conceded that, having regard to the relatively short time involved and to the continuing trading relationship between the parties, in the event that interest was awarded it should, up to the date of the award, be simple interest.

    (The rate adopted here is notional, for purposes of illustration only.)

(x) there are no submitted grounds for interest being non-continuous[1];

(xi) the basis and amount are shown in the award.

An example relating to the present case might be (here using a notional percentage (5%) simply to enable a cash figure to be computed and shown):

---

[1] For illustration relating to an 'intervening event', see 9.3.9.1.

Interest

41   I now have to consider interest on the amount I have determined as due to the Claimant. Contractual interest is not applicable. It is, therefore, a matter for my discretion under section 49 of the Arbitration Act 1996.

42   The amounts in question were invoiced (although in greater amounts than I have determined) on 18th February [*year*], a few days after completion of the painting in issue. There was contractual provision for payment against invoices to be made within 28 days. The Respondent has contended that I should not award interest, arguing that the Respondent did not pay on time simply because of what he contended to be grossly excessive charges. I have rejected that contention. Additionally, the Claimant's invoiced amount, though high, was not excessive, nor had it been so would that have been good reason for failing to pay a reasonable amount against the invoice. Regarding interest, then, I shall exercise my discretion in favour of the Claimant, the period to run from 28 days after the date of the invoice (and thus from 18th March [*year*]). As to the rate, having seen credible evidence (which was not disputed) that the Claimant's average bank balances over the preceding years were generally 'neutral', with deposits and borrowing being approximately in equilibrium, I shall adopt a rate slightly above the average base rate over the relevant period (i.e. from 18th March [*year*] to the date of this award). The Claimant at the hearing conceded that in all the circumstances, if I were to allow interest this could be on a simple interest basis. I therefore adopt 5% simple interest.

43   **In the exercise of my discretion, therefore, I shall award on the sum of £17,100.00 simple interest at 5% per annum, from 18th March [*year*] until the date of this my award, which interest I compute at £269.38.**

5.3.9.2   Post-award interest   [F(a)3;(b)]

Regarding post-award interest ('tasks' (xii) and (xiii)), the rate applied will commonly be higher than that in the award. The party due to pay has prior warning of what will be faced in the event of late payment and can act accordingly.

Factors affecting choice of date from which post-award interest is to run are discussed in Chapter 4 Section F. If either kind of interest is awarded on what could appear to be an unusual basis, the reason for so doing should be given. If the provisions of any other relevant statute are applied, that should be so stated (see reference to post-award interest in 4.3.3.1(5) and in the tasks in 4.3.3.2, at (xii)).

Brief further reference is made at 9.3.11.2 to post-award interest and the commencement date for such interest.

For present purposes, provision might be:

> 44   **As to post-award interest, again in the exercise of my discretion under section 49, I shall provide for compound interest to be payable at [rate] % per annum compounded with quarterly rests from the date of this award until the date of payment.**

Section G

5.3.10   *Costs* [G]

5.3.10.1   Award of costs [G(a)–(l)]

Factors affecting the award, or not, of recoverable costs are discussed in Chapter 4 Section G. The enumerated notes below refer to the 'tasks' (i)–(xiv) in that section rather than the checklist as such. See 6.4 for the effect of a counterclaim and 9.3.10.1 for effect of a rejected offer of settlement.

Unless to be reserved to a later award, the arbitrator should now consider whether or not a successful party (claimant or counterclaimant) is entitled to reimbursement by the other party of its recoverable costs or any part thereof. If, exercising discretion under the exception in section 61(2), the arbitrator departs from the normal rule of 'costs follow the event', the reasons for so doing should be stated.

In this present illustration (but these matters only need to be recited if and to the extent that they affect the decision):

(i)   there were no constraints upon the arbitrator's powers in this regard;

(ii)   the parties had requested that liability for costs be determined by this award, leaving quantum of costs for agreement or further award;

(iii) there was no valid prior agreement on who should pay costs;

(iv)   as there was no counterclaim the matter of set-off does not arise;

(v)/

(vii) the only matters affecting the principle that 'costs follow the event' were the two orders recited at award paragraphs 9 and 11 whereby the respondent was made responsible for certain costs, but as the claimant is successful on the substantive issues those sanctions are merged in the respondent's overall liability for costs; there were no grounds for allowing only a proportion of the claimant's costs;

(vi)   there was no application to vary from the default provisions of section 63(5);

(viii) no limit had been imposed under section 65;

(ix)   no security for costs had been ordered under section 38(3);

(x)   there has been no provisional order under section 39 relating to costs;

(xi)   there had been no offer of settlement of the substantive issues;

(xii) there were no grounds whereby the basis of recoverable costs should be other than the default position under section 63(5); as to 'Fit for counsel', both sides had been represented by counsel and no issue had arisen in that respect (and this is therefore referred to in the 'operative' section only);

(xiii) as indicated, and by agreement, quantum only is to be reserved to a further award (if not agreed).

(xiv) there is no reason for any apportionment of liability for arbitrator's fees and expenses; in this illustration, interest is applied to any inter-party reimbursement (but see 4.3.3.1(4) para 2).

In these circumstances, the award might include:

> **Costs**
>
> 45   I find no reason to depart from the normal rule. The Claimant was successful in its claims with only a modest reduction in quantum and the only 'costs in any event' orders were against the Respondent. In consequence the Claimant is entitled to its recoverable costs. I have been asked to reserve quantum.
>
> 46   **I determine, therefore, that liability for costs (including my fees and expenses) shall follow the event**.

The settling of the quantum of recoverable costs is, by that decision and unless the parties otherwise agree, squarely on the arbitrator's shoulders. It can be dealt with at this stage only in the now relatively unusual situation that submissions on quantum of costs have already been made, or the amounts are fixed under some applicable scheme or rules. It is more likely in present circumstances that the determination of the amount of recoverable costs is reserved to a further award, as has been done here. Should that need to determine the amount arise, there would in effect be an arbitration on costs (usually within the ambit of the existing arbitration agreement – and hence the specific reservation of that function (award paragraphs 48(d) and 50 below)).

Section H

5.3.11   *Operative section (award)*   [H]

In a complex case, or one with many issues, and if not already effectively summarised by what has previously been set down, it can be useful, in advance of this heading, to summarise the various conclusions before setting down the operative part of the award. That is particularly so if interest awarded varies from one element to another (see at 5.3.9.1). Such a summary is not necessary in this illustration.

It is then appropriate to have a section specifically headed 'AWARD', which effectively gives such 'instructions' as are necessary to give effect to the decisions which have been made (see Sections D–G above), thus:

AWARD

5.3.11.1 Introduction [H(a)]

Some arbitrators feel a need to recite the fact of having read, considered and taken note of the written submissions etc., the evidence and the addresses of advocates. In some awards that is self-evident from the manner in which the arbitrator's consideration of the issues has been set out. If it is not, the arbitrator might wish to insert something like (therein identifying the nature of the submissions, evidence, and addresses if any):

> 47 I have read, considered and taken note of the statements of case, the written and oral evidence and the addresses by Counsel.

5.3.11.2 Type of award [H(b)1,2]

This present illustration relates to a monetary award where there was no counterclaim. Principal and interest can be given as separate items, or might be combined as two parts of one item. If, as here, any reimbursement of VAT is then to be dealt with by a declaratory award referring solely to the principal monetary award, it is convenient to show the principal figure to which VAT is to be applied (see award paragraph 48(c) below). See 6.4 for the effect of set-off and counterclaim respectively. Non-monetary awards are illustrated in Chapters 6–8.

5.3.11.3 Conditions or terms [H(c)]

Payment under an award might sometimes be conditional on some event, for instance agreed or certified completion of the work or services in question. Any such condition should be incorporated here. None apply in this illustration.

5.3.11.4 Time for performance [H(d)]

If not inherent in any condition under H(c) above, the award must state when (or by when) payment is to be made (or, if a non-monetary award, when the consequence of the award is to be put into effect). In this present illustration that is to be within a stated time of the award being taken up. For further comment see 9.3.11.1.

As to [H(b)] and [H(d)] above, the entry in these circumstances (there being no counterclaim) might be:

> 48 I AWARD AND DIRECT THAT, in full and final settlement of all claims in this arbitration:
>
> (a) the Respondent shall within [*number*] days after that when either party shall have taken up this award pay to the Claimant:
>
> (i) the sum of £17,100.00 (seventeen thousand one hundred pounds); *and*
>
> (ii) interest as computed at paragraph 43 above and amounting to £269.38 (two hundred and sixty nine pounds and thirty eight pence); *and*

5.3.11.5 Post-award interest   [H(e)]

Reflecting the decisions relating to basis and date already made (Section F), the item might continue:

> (b) the amounts awarded at paragraph 48(a)(i) and (ii) above (or, in the event of part payment by the due date, the outstanding balance thereof) shall be subject to the addition of compound interest at [*rate*] per cent with quarterly rests from the date of this award up to the date of payment; *and*

See briefly at 9.3.11.2 regarding starting date.

5.3.11.6 Value added tax   [H(f)]

In this instance all that is needed is a simple declaration (but see brief reference at end of 4.3.3.1(1) regarding interest, and 9.3.8 for further illustrations), so that item 48 could continue:

> (c) the Respondent shall within [*number*] days of receipt by it of an appropriate VAT invoice from the Claimant pay to the Claimant such Value Added Tax, if any, as is properly chargeable under the applicable VAT Regulations on the principal sum awarded at paragraph 48(a)(i) above, {[2]and interest thereon at [*rate, simple or compound, period of rests*] from the last date for receipt thereof until the date of payment by the Respondent}, *and*

[2] As will be appreciated from the comments on VAT at 4.3.2, a variety of matters could affect this interest, and its inclusion or not. It would only be included (and appropriately phrased) if it had been the subject of submissions and had been dealt with in the preceding analysis section.

5.3.11.7 Costs [H(g)]

Reflecting the decisions made in Section G, this item might further continue (assuming that the *amount* of costs is being left over for party agreement or subsequent determination):

> (d) the Respondent shall pay the Claimant's recoverable costs, together with interest thereon at [*rate, simple or compound, period of rests*] from the date of this award until payment, the amount of such costs and interest thereon (failing agreement) to be determined by me {[3]on the basis described at sub-section 63(5)(a) and (b) of the Arbitration Act 1996} by my further award; *and*

That could continue, making provision for the possibility that a party not liable for the arbitrator's fees has in the first instance paid all or part of them[4]:

> (e) the Respondent shall also pay and bear my fees and expenses in respect of this my First Award, which fees and expenses I determine at £[*amount; figures and words*], plus £[*amount*] Value Added Tax, thus totalling £[*amount; figures and words*], and to the extent that the Claimant has paid any part thereof the Respondent shall forthwith reimburse him with that amount {[4]and interest thereon at [*rate, simple or compound, period of rests*] from the date of the Claimant's payment until the date of such reimbursement}.

See 9.3.11.3 regarding release or disposal of any security for costs.

Then, the following item is simply a shorthand term to indicate (for instance to the court in the event that the amount of recoverable costs is to be so determined under section 63(4)) that the arbitrator has held that the case was such as warranted the involvement of Counsel (see note at (xii) in Section G):

> 49 Fit for Counsel.

Section I

5.3.12 *Reserved matters* [I]

If the award relates only to certain issues or, for instance, leaves costs to be dealt with in another award, or the like, it is desirable to reinforce whatever qualification might have been incorporated

---

[3] See 4.3.4.1(3) regarding any higher, or other, basis.
[4] But regarding interest, see 4.3.3.1(4) paragraph 2.

in the title (e.g. 'Final Award except as to costs') by specifically reserving that/those remaining function(s). An example here would be:

> 50   I RESERVE to my further or final award determination, if not agreed, of the amount of recoverable costs and any interest thereon.

Some arbitrators append to this kind of provision what is in effect an administrative note, such as:

> (for which purpose I shall give my further directions upon the application of either party).

whilst others deal with that outside the award, by Order or letter.

Section J

5.3.13   *Signature and formalities*   [J]

Having completed the draft award and before signing it, the arbitrator should have full regard to the note 'Final check' at 3.3.14.

Subject to party agreement otherwise and the provisions of section 52, the award must be signed by the arbitrator(s) and it must state when the award is made. It has been traditional also to state where the award is made – and see section 53 regarding the default position. If the place where the award is made is located within the designated juridical seat, there should be no problem in stating both the seat and the place where the award was made (as at {*} in the illustration below).

If, whilst 'made' within the designated seat, the award is perchance signed within another, then, if the award has known or potential international implications, it is necessary to avoid any foreign court being misled as to the applicable seat. Some commentators advise for all awards, domestic or otherwise, that *only* the seat should be stated, relying on section 53 to define where the award is *deemed* to have been made. (Section 53, however, has an 'Unless otherwise agreed by the parties' provision.) If the seat has not been recited earlier in the award it can be conveniently stated here.

The effect and impact of the date is discussed at 1.3.1.5.

If the arbitrator's address has been recited earlier there is no need to repeat it here unless the award is made from a different (e.g. office) address, but it may be. If not recited earlier it should be set out here, if only as a means of positive identification. To have the signature witnessed is at least a prudent step (see 1.3.1.1).

It is prudent to ensure that this 'attestation' of the award is positively linked to the foregoing award content, usually by it appearing on the same page as the concluding paragraphs of the award. If reasons appear as an appendix, there should be an adequate attestation to link them positively to the award as such (and see 9.3.7.2).

A typical attestation is commonly something like:

This my First Award made and published by me {*in Nevekirk, England} on [*date*].

[*Signed*].......... *Jonathan Fairley* .....

Jonathan Fairley, Arbitrator

in the presence of: *Anne Gee* .... [Signature of witness]

of ... *The Lodge, Weiring Common, Weirshire.* [address]

...... *Secretary.*........................... [occupation]

Reminder

The foregoing illustration is simply a starting point, an illustration stemming from *one* form of checklist. Subject to compliance with the requirements of a valid enforceable award, the format, sequence and form of narrative are matters for each individual arbitrator and are also dependent upon the circumstances of the particular case.

Part D
Illustrations

6 Substantive awards

Monetary awards – Declaratory awards – Performance awards – Injunctive awards – Rectificative awards

6.1 Introduction

Comments and illustrations relate to English law and are on the basis that, unless stated otherwise, all 'default' provisions of the 1996 Act apply without alteration, and there are only two parties.

The illustrations in Chapters 6–9 relate in the main to the same basic scenario as set out at 5.2, but adapted on each occasion so as to provide an extended or amended background for each succeeding illustration. See 5.1, 'Introduction to the illustrations', for general explanations.

The first illustration in the present chapter, at 6.3.1, is the monetary award synthesised in Chapter 5 and repeated, without the intervening narrative, as Appendix 3. The illustration at 6.3.2 relates to the same scenario, but assuming that the claim had failed.

The next adaptation of that award, relating to the same case but subject to a counterclaim which constitutes a set-off, is shown comprehensively at 6.4.1 by setting out or indicating the relevant changes. That is followed by further monetary illustrations at 6.4.2 (a true counterclaim) and 6.4.3 (a mixed variant), before addressing other types of substantive awards.

The further illustrations in this and later chapters concentrate upon the impact of the changed scenario upon the *operative* section of each award, with a minimum of narrative explanation or illustration of related award content. Incidental consequential changes to the underlying content should be self-evident to the reader and it would be unnecessarily space-consuming to set out each and every change. Whilst in practice there might well have been further interlocutory matters to recite as a consequence of each change to the scenario, such changes are ignored for present purposes (but see Chapter 9). In the various insertions, paragraph numbers marked '/' (e.g. 18/) indicate that such paragraphs could be inserted, in sequence, after that numbered paragraph in the Chapter 5 illustration. Where an illustration does not relate directly to an award paragraph in an earlier illustration, its suggested approximate position is shown by the relevant checklist reference.

The various illustrations which follow should be read in conjunction with Chapter 5, for some of the non-illustrated narrative comments in that synthesis will come into play or have relevance to what follows in Chapters 6–9. As in Chapter 5, for ease of cross-reference the sub-headings (and here parts of the explanatory narrative) show the checklist references, in square brackets, in addition to the decimal enumeration. However, as in much of Chapters 6–9 only some of the checklist items arise, specific cross-referencing within and between these chapters is by the principal decimal enumeration. The term 'paragraph' used in the following narrative refers to the numbered paragraphs in the award synthesised in Chapter 5.

The non-monetary awards illustrated would still need to include or to reserve the decision on costs and any interest thereon (see 6.4.2.6, and also see 7.5 regarding interest on recoverable costs), but that is not illustrated again in those non-monetary awards.

6.2 A monetary award

Solely for convenience in illustrating various basic forms of monetary award, each is here treated separately (at 6.3 and 6.4), before proceeding to other forms of award.

6.3 A monetary award without any counterclaim

6.3.1 Where the claim is successful

As indicated above, the relevant illustration of a monetary award with no counterclaim is set out within Chapter 5, repeated without the intervening narrative as Appendix 3.

6.3.1.1 Summary [D(d)]

Diverting briefly from the basic scenario – a party, whilst generally successful, might have succeeded on some issues but not on others. Reference has been made earlier (e.g. at 5.3.7.4) to the effect of various factors upon the nature and location of any summary of the arbitrator's decisions. The following is a simple illustration of a summary in a situation where a claimant was successful on five out of six monetary issues. For this illustration it is assumed that interest on all issues would relate to the same period and at the same rates.

Summary

[D(d)]

[*no.*] For convenience, my decision on Issues 1 to 6 are summarised below:

Issue 1 [*description*]	£17,260.00
Issue 2 [*description*]	2,130.00
Issue 3 [*description*]	Fails
Issue 4 [*description*]	7,390.00
Issue 5 [*description*]	3,120.00
Issue 6 [*description*]	11,550.00
Total, before interest	**£41,450.00**

I consider the matter of interest later, at paragraph [*number*].

The consideration of interest might, depending upon circumstances, follow the general pattern shown in Chapter 5 at award paragraphs 41–43, but (to suit this particular scenario only) incorporating a reference that the commencing date is the same in respect of interest on all issues. If not obvious, the reason for that being so should be set out. The question of whether the claimant should be awarded all of its recoverable costs in this instance, or some lesser proportion, would depend upon how significant the failed Issue 3 was, in terms of amount claimed and time expended upon it.

6.3.2 *Where the claim fails*

The illustration referred to in the preceding paragraph relates to a situation where the claimant was successful or substantially so. Had, instead, the decisions on sub-issues (a) and (b) been such that the claimant was held to have included the work in question in the contract sum:

(a) there would have been no need in Section D to consider the other sub-issues;

(b) by award paragraph 39 the arbitrator would have determined that the claimant is not entitled to additional payment for executing the work in dispute; *and*

(c) consideration of VAT under Section E and of interest under Section F would not arise.

In Section G dealing with liability for costs, whilst in the circumstances there would still be no reason to depart from the normal rule, the claimant was unsuccessful and would (subject to any 'costs in any event' orders) be liable for costs. The only costs in any event order (see award paragraph 11) had been against the respondent, so *that* part of the costs would be the liability of the respondent. In consequence, paragraphs 45 and 46 might instead be:

Costs

45 I find no reason to depart from the normal rule. The Claimant was unsuccessful in its claim, but by my Orders Nos [number] and [number] the Respondent was made responsible for costs of and resulting from those orders. I hereby confirm and incorporate those Orders as part of my award. In consequence the Respondent is entitled to its recoverable costs other than those relating to those two Orders; the Claimant is entitled to its costs in respect of those two Orders. I have been asked to reserve the determination of quantum of costs. As to my fees and expenses, liability follows that same principle and I compute the proportion thereof occasioned by those two Orders to be 5% (five per cent).

46 **I determine, therefore, that liability for costs (including my own fees and expenses) shall follow the event subject to the foregoing.**

Effect is given to that determination at award paragraphs 48(b) and (c) below.

In the operative section H (AWARD) these various decisions could replace award paragraph 48 (47 and 49 being unchanged) (but with regard to 48(b)(i) and (ii); see comment at 4.3.4.1(5) concerning a single overall proportionate award of costs, which can also apply here):

AWARD

48 I AWARD AND DIRECT THAT, in full and final settlement of all claims in this arbitration:

(a) the Claimant's claim fails; *and*

(b) as to costs:

(i) the Claimant shall pay the Respondent's recoverable costs other than those of and resulting from my Orders Nos [number] and [number], together with interest on the net amount thereof at [*rate, simple or compound, period of rests*] from the date of this award until payment; *and*

(ii) the Respondent shall pay the Claimant's recoverable costs of and resulting from those two Orders, together with interest thereon at [*rate, simple or compound, period of rests*] from the date of this award until payment;

the respective amounts thereof {[1]on the basis described at sub-section 63(5)(a) of the Arbitration Act 1996} and any interest thereon (failing agreement) to be determined by me by my further award; *and*

---

[1] See 4.3.4.1(3) regarding any higher or other basis.

> (c) the Claimant shall pay and bear 95% and the Respondent 5% of my fees and expenses in respect of this my First Award, which fees and expenses I determine at £[*amount; figures and words*], plus £[*amount*] Value Added Tax, thus totalling £[*amount; figures and words*], and to the extent that either party has paid any part thereof in excess of those proportions the other party shall forthwith reimburse such party with the amount of that excess {[2]and interest thereon at [*rate, simple or compound, period of rests*] from the date of payment until the date of such reimbursement}.

As to Sections I and J, the preceding changes would not necessitate any changes to those sections.

6.4 A monetary award with a counterclaim

The effect of a counterclaim is illustrated below:

(a)   where such cross-claim constitutes a set-off;
(b)   where it does not, then
(c)   with further variants.

6.4.1 Counterclaim which constitutes a set-off

Further scenario

Shortly before the parties had concurred in the appointment of an arbitrator on the painting issue, STL complained that decoration on some wall areas was flaking and peeling. Other than 'noting' the allegation, TGP took no action. With new productions looming in which the walls in question would be visible, the matter was urgent. STL employed its own stage staff, who were accustomed to painting stage sets, working after normal working hours to scrape down and redecorate the walls in question, having given TGP notice of that intention prior to starting. On the basis of staff time recorded on the theatre's normal costing system, the total cost excluding any 'profit' element was £1,100.00. STL told TGP that it required reimbursement of that net amount, foregoing any mark-up. As the whole matter of the decoration was about to go to arbitration, although no arbitrator had at that time been appointed, the parties agreed that this related matter should also be referred to the same arbitrator as a counterclaim in the same reference.

Whilst, as indicated, subsequent illustrations will not be in this amount of detail, the overall consequent effect upon the award illustrated in Chapter 5 is shown below.

---

[2]   See 4.3.3.1(4) regarding pre-award interest on costs.

6.4.1.1   Heading and preamble [A]

The second paragraph of the preamble would be expanded by a suitably condensed version of the further scenario shown above.

6.4.1.2   Dispute/matters in issue [B1(g);(h)]

Paragraph 5 (B1(g) and (h)) would be expanded, first inserting a sub-heading 'The Claimant's claim' after the first sentence. After sub-item (i) the counterclaim would be incorporated, here as part of paragraph 5 (but it could be a separately numbered item), possibly thus:

> **The Respondent's counterclaim**
>
> The Respondent asserts that, after completion of the contract works, but before the re-opening, it found that areas of the new decoration to walls were flaking and peeling and that:
>
> (a) the Claimant having taken no action after it was informed of the problem, in view of the urgency the Respondent, having given the Claimant notice of that intention, made good the defect using its own stage staff working after normal hours; *and*
>
> (b) on the basis of its normal costing system, excluding any addition for profit or 'mark-up', the cost of that remedial work amounted to £1,100.00; *and*
>
> (c) the Respondent having informed the Claimant that it required reimbursement of that net amount from the Claimant, the Claimant demurring, it was agreed that this further issue be treated as a counterclaim in this arbitration (in respect of which no appointment had at that date been made); *and*
>
> (d) the Respondent now claims by way of counterclaim:
>
>   (i)   the said sum of £1,100.00;
>
>   (ii) interest under the said section 49; *and*
>
>   (iii) costs of the counterclaim;
>
> and the Claimant contends that:
>
> (e) whilst it accepts that some paint-work did fail and that the sum counterclaimed is fair for the area of walls said to have been re-treated, it alleges that that area is excessive, not having been necessary to that extent; *and*
>
> (f) the Claimant claims costs of the counterclaim.

Followed by the original last line of paragraph 5.

Had the matter of appointment not conveniently been merged in the paragraph set out above, it could be dealt with by extension of paragraph 6 under B1(i).

6.4.1.3   Issues in this award   [B2(g)]

Paragraph 8 (B2(g)) would be expanded, first inserting a sub-heading at the beginning: 'The Claim'. After sub-item (e), the counterclaim issues would be incorporated, possibly thus:

> **The Counterclaim**
>
> The Counterclaim issues to be determined herein are:
>
> (a) what area of walls was it necessary to redecorate?
>
> (b) what was a fair deduction to be made by the Respondent in respect thereof?

6.4.1.4   Inspections   [B3(g)]

Paragraph 16 (B3(g)) would be expanded to incorporate a further aspect of the inspection, with the result of the inspection being given here or later under C(d)2:

> Additionally, my attention was drawn to the redecorated walls, the perceptible margins of and extent of recent flaking or peeling, and the extent of redecoration relative to those defects – and I was asked by the parties to take account of what I saw.
>
> **In that connection, I saw that the area of the most recent redecoration extended beyond the recognisable boundaries of the said flaking or peeling to the extent of ending at internal or external angles of the walls, masking any possible slight mismatch of tone. I made those present aware of what I had seen in that respect.**

6.4.1.5   Contentions re counterclaim   [C(a)]

In paragraphs 17 and 18 (C(a)) (or, suitably adjusted, as a general subheading) there would be added: 'of the Claim'.

There would then be added, after paragraph 18, numbered paragraphs such as:

> 18/   **Relating to issues (a) and (b) of the counterclaim:**
>
> **The Respondent** contends that:
>
> (a) the area of wall redecoration was limited to that reasonably necessary; *and*
>
> (b) the amount of the counterclaim was a fair and proper amount to deduct or recover from the Claimant.

The Claimant contends that:

(a) there was no need to redecorate to the extent which was done; but the Claimant is content to abide by my conclusions derived from the inspection; *and*

(b) a fair amount would be pro rata to the area determined by me, relative to the area repainted.

6.4.1.6 Analysis etc./decisions/counterclaim or set-off [D(c)]

In this further illustration, the counterclaim has been kept simple so as more easily to concentrate upon the 'set-off' point.

After the heading 'My consideration of the issues' (D(c)), there would be inserted a sub-heading: 'The Claim'. In the decision on the claim, at paragraph 39, would be added 'subject to the counterclaim' and that would be followed, under a sub-heading, by numbered paragraphs such as:

The Counterclaim

Issue (a) : What area of walls was it necessary to redecorate?

39/ The Claimant has said that it is content to abide by my conclusion decided from my inspection. At the inspection I drew attention to the fact that the redecoration extended beyond the defects only sufficiently, i.e. up to wall angles, as to mask any possible slight difference of tone due to extra coats of paint. That extension was of very modest area and was reasonable in the circumstances.

39/ **I determine that the area of walls which it was necessary to redecorate was that in fact redecorated.**

Issue (b) : What was a fair deduction to be made by the Respondent in respect thereof?

39/ The Claimant accepts that the sum counterclaimed is fair for the area in fact redecorated but contends that it should be reduced pro-rata to the area determined by me. As I have determined that area to be as executed, the amount remains as claimed.

39/ **I determine that a fair deduction to be made by the Respondent is, therefore, £1,100.00.**

There would then follow (D(c)6) the consideration of whether that sum constituted a true counterclaim or a set-off. Either here, or elsewhere if more convenient to the flow of the award (e.g. in Section G: Costs), the following might be inserted, under a suitable sub-heading:

Counterclaim or set-off

39/ On this point, Mr Blackwelle suggested that the facts speak for themselves and was content to leave the decision with me without specific submissions. Ms Wellbie concurred.

39/ The counterclaim stems directly from the claim and is inextricably connected with it. Its effect if successful, as here, is to reduce the amount otherwise found as due to the Claimant.

39/ **I determine, therefore, that the counterclaim amounts to a set-off and that when later considering liability for costs there is one 'event'.**

The position in the award of any summary of monetary decisions (D(d)), if needed, will depend upon circumstances such as (see 5.3.9.1):

(a)   whether there is a list of issues (claim and/or counterclaim) which need to be totalled before considering interest;
(b)   whether interest needs to be considered separately on claim and counterclaim.

In this present instance there is only one sum on each of the claim and counterclaim, so any summary can conveniently be left until interest has been considered (see [F(b)] later).

6.4.1.7   VAT implications   [E(a)–(c)]
There are no VAT implications on this counterclaim, but to avoid uncertainty it would be prudent to recite that here, adding to paragraph 40 (E(a)-(c)):

As to the counterclaim, both Counsel confirmed that no VAT implications arise, so no determination thereon is required of me.

6.4.1.8   Interest   [F(a);(b)]
In paragraph 43 (F(a) and (b)), there would be inserted after 'I shall award': 'to the Claimant'.

Regarding interest, the respondent has claimed interest under section 49 of the Arbitration Act 1996. Once again, contractual interest is inapplicable. The respondent has already paid for remedial work, so interest is applicable to the claim and set-off separately. (See under 4.3.3.1(3) above for a situation where set-off has been computed as at the date of the award.) In the present instance, the date when the respondent incurred the cost of redecoration was, perchance, within a day or so of the date when the claimant's invoice was due for payment. However, the rate of interest awarded might differ (see award paragraph 42). For the purposes of this illustration, it is assumed that the *claimant was aware* that the respondent was relying throughout on bank finances and would suffer bank interest charges at (say for the purpose of illustration only) 9% with quarterly rests. In those circumstances the effect on the claim and counterclaim need to be computed separately, and this (after this insertion following paragraph 43) is also a convenient place to summarise figures:

43/   Turning to the matter of interest on the counterclaim, also claimed under section 49, the Respondent incurred the cost of redecoration on 25th March [*year*] and is entitled to interest from that date. As to rate, Ms Wellbie having admitted that the Claimant was aware from the outset that the Respondent operated on borrowed finance, it is appropriate that I allow interest at the borrowing rate incurred by the Respondent, viz 9%, again as simple interest.

43/   **In the exercise of my discretion, therefore, I shall award to the Respondent simple interest at 9% per annum on the sum of £1,100.00, from 25th March [*year*] until the date of this my award, which I compute at £29.29.**

Summary

43/   The amalgamation of these figures gives:

Awarded to Claimant		£17,100.00
Interest		269.38
		17,369.38
Less: Set-off	£1,100.00	
Interest	29.29	
		1,129.29
Net amount to Claimant:		**£16,240.09**

6.4.1.9   Costs – single event   [G(c)]

As in this instance the counterclaim constitutes a set-off, there is only one event, on which the claimant is successful even if in a lesser amount than claimed. In consequence (unless circumstances were such as to warrant the application of some degree of proportionality, awarding the claimant less than its total recoverable costs), the position on liability for costs remains unaltered from that shown in Chapter 5.

Paragraph 45 (G(a)–(k)) could simply have inserted after 'reduction in quantum' '(in general and by way of set-off)'.

6.4.1.10   Operative section   [H]

As the successful counterclaim has been set off against the successful claim, it was convenient to amalgamate the interest figures in the same summary. In consequence the substantive sum and interest can be awarded as a single sum. However, the figure to which the declaration in paragraph 48 regarding any applicable VAT relates is that shown in the body of the Chapter 5 award at paragraph 43, relating to the claimant's claim. (The value of the VAT taxable supply, had it not been defective, would remain unaltered, i.e. as awarded at £17,100.00).

As a number of minor changes to phraseology are necessitated, the following could replace 48(a)–(c), leaving (d) and (e) unaltered:

> 48   I AWARD AND DIRECT THAT, in full and final settlement of all claims and counterclaims in this arbitration:
>
> (a)   the Respondent shall within [*number*] days after that when either party shall have taken up this award pay to the Claimant the sum of £16,240.09 (sixteen thousand two hundred and forty pounds and nine pence) inclusive of interest to the date of this award; *and*
>
> (b)   the amount awarded at 48(a) above (or, in the event of part payment by the due date, the outstanding balance thereof) shall be subject to the addition of compound interest at [*rate*] per cent with quarterly rests from the date of this award up to the date of payment, *and*
>
> (c)   the Respondent shall within [*number*] days of receipt by it of an appropriate VAT invoice from the Claimant pay to the Claimant such Value Added Tax, if any, as is properly chargeable under the applicable VAT Regulations on the sum of £17,100.00 determined above, {[3] and interest thereon at [*rate, simple or compound, period of rests*] from the last date for receipt thereof until the date of payment by the Respondent}; *and*

That could be followed by the original (d) and (e), and paragraphs 49 and 50. In turn, there could follow any reserved matters, and signature and formalities, which in the present case remain unaltered.

6.4.2   *Counterclaim which does* not *constitute a set-off*

Further scenario

During the period when the sets were being delivered, but unconnected with this contract as such, a driver employed by TGP, but without authority, drove his low-loader into the theatre car park simply to pick up a magazine from a TGP operative friend. When turning, the 'unauthorised' driver caused damage to the theatre flank wall.

---

[3]   See comment at 5.3.11.6.

STL has had that wall rebuilt at a cost of £4,745.00 plus £830.76 VAT. STL treated the VAT as input tax for VAT purposes and claimed only the £4,745.00 from TGP. TGP accepted responsibility for its employee's tortious action but rejected the amount of the claim. Later, but prior to the joint invitation which initiated the 'painting' arbitration, both parties agreed that the matter be dealt with within that arbitration.

6.4.2.1 Background and submissions   [B,C]
The award would take a similar form to that illustrated at 6.4.1 above, amended to suit this scenario, including the VAT situation. *Only those paragraphs which are essentially different from the last illustration are shown below.* For the purpose of illustration here (which is principally to develop the 'set-off or not' matter) it is assumed that STL were able to prove that the amount paid, and claimed, was fair and reasonable – and succeeded on that point.

6.4.2.2 Counterclaim or set-off   [D(c)6]
Under the (paragraph 39/) sub-heading 'Counterclaim or set-off' (D(c)(vi)), the second and third paragraphs shown would be replaced by:

> 39/ The driver and vehicle concerned, and the purpose of the visit, had no connection with the STL/TGP contract. The counterclaim does not stem from nor have any causal connection with the Claimant's claim. It has an independent existence.
>
> 39/ **I determine, therefore, that the counterclaim is not a set-off and that it constitutes a 'separate event' and will be so treated when considering liability for costs.**

6.4.2.3 Interest   [F]
Regarding Section F, Interest, assuming that the cost of repair was (coincidentally) incurred on 25th March [*year*], the first paragraph to be inserted would be similar to paragraph 43/ in the illustration at 6.4.1.8. The second paragraph would also be the same, but replacing the figures with £4,745.00 and £126.36.

In considering the summary (see the last paragraph 43/ inserted in 6.4.1.8), it is necessary to have regard to:

(a) whilst the award (Section H) could show a single net figure (claim less counterclaim) it can be less confusing if the two are shown separately, particularly if there are differing VAT implications; *and*

(b) VAT relates, for example, to the claim but not to the counterclaim; *and*

(c) there are two events, thus influencing liability for recoverable costs;

so that the figures are best kept separate at this summary stage. The summary could therefore be:

Summary

43/ The respective sums to be awarded are, therefore:

The Claimant's claim:

Awarded	£17,100.00
Interest	269.38
	£17,369.38

The Respondent's counterclaim:

Awarded	£4,745.00
Interest	126.36
	£4,871.36

6.4.2.4 Costs – separate events [G(c)]

Considering costs, as contrasted with the illustration at 6.4.1, there are two events here. See Chapter 4 at 4.3.4.1(5) for alternative ways of dealing with this. For the present illustration the costs are awarded separately. Instead of the paragraphs 45 and 46 in Chapter 5 there could be:

Costs

45 The Claimant was substantially successful in its claims and the Respondent was successful in its counterclaim. There were two 'costs in any event' Orders against the Respondent, both of which related to the counterclaim. I hereby incorporate and confirm those Orders as part of my award. In consequence, the Claimant is entitled to its recoverable costs of the claim and the Respondent is entitled to its recoverable costs of the counterclaim *other than* those costs of and resulting from my Orders Nos [*number*] and [*number*]; and the Respondent shall pay the Claimant's costs of and resulting from those Orders.

46 **I determine, therefore, that liability for costs (including my own fees and expenses) shall, subject to the foregoing, follow the two events.**

See also award paragraph 48(d) below.

As with a claim, in some circumstances a counterclaim, too, might be only partially successful, warranting an award for only part of the costs of the counterclaim.

6.4.2.5 Operative section [H]

Regarding the substantive claims, some arbitrators prefer to award a net sum (in the present case to the Claimant), so that no further calculation needs to be done by the parties. Others feel that the separate decisions (claim and counterclaim) should appear in the operative section as such, in each case showing both the substantive sum and the interest. The format adopted here in effect merges the two approaches. As in this instance the sum (£17,100.00) to which VAT might relate (in whole or part) appears in this award paragraph, there is no need to refer to a previous paragraph for identification of that component (as was necessary in the illustration at 6.4.1.10).

Relating to the arbitrator's fees, in this illustration the arbitrator, having heard relevant submissions, has been able to assess the proportion which relates to the parties' relative liability for costs, having regard to (a) his time involved on the counterclaim and (b) that occasioned by the two Orders. In doing so he must act judicially.

Instead of the illustration of paragraph 48 at 6.4.1.10, there could appear:

48 I AWARD AND DIRECT THAT, in full and final settlement of all claims and counterclaims in this arbitration:

(a) the Respondent shall within [*number*] days after that when either party shall have taken up this award pay to the Claimant the sum of £12,498.02 (twelve thousand four hundred and ninety eight pounds and two pence), made up of: on the claim, £17,100.00 plus £269.38 interest, less, on the counterclaim, £4,745.00 plus £126.36 interest; *and*

(b) the amount awarded at 48(a) above (or, in the event of part payment by the due date, the outstanding balance thereof) shall be subject to the addition of compound interest at [*rate*] per cent with quarterly rests from the date of this award up to the date of payment, *and*

(c) the Respondent shall within [*number*] days of receipt by it of an appropriate VAT invoice from the Claimant pay to the Claimant such Value Added Tax, if any, as is properly chargeable under the applicable VAT Regulations on the principal sum awarded to the Claimant and identified in paragraph 48(a), viz £17,100.00 [4 *and provision for interest if applicable*], *and*

---

4 See comment at 5.3.11.6.

(d) (i) the Respondent shall pay the Claimant's recoverable costs of the claim;

    (ii) the Claimant shall pay the Respondent's recoverable costs of the counterclaim *other than* those costs of and resulting from my Orders Nos [*number*] and [*number*];

    (iii) the Respondent shall pay the Claimant's costs of and resulting from those two Orders; *and*

    (iv) the respective amounts thereof shall be paid together with interest thereon at [*rate, simple or compound, period of rests*] from the date of this award until payment;

the amount thereof and any interest thereon (failing agreement) to be determined by me [*state basis*] by my further award; *and*

(e) the Claimant shall pay and bear 85% and the Respondent 15% of my fees and expenses in respect of this my First Award, which fees and expenses I determine at £[*amount; figures and words*], plus £[*amount*] Value Added Tax, thus totalling £[*amount; figures and words*], and to the extent that either party has paid any part thereof in excess of those proportions the other party shall forthwith reimburse such party with the amount of that excess [5].

Paragraph 49 would remain.

6.4.2.6   Reserved matters [I]

The arbitrator has to consider what matters, if any, should be reserved in this particular case. As the parties might dispute just what *elements* of costs pertain to the facets 48(d)(i)–(iii) above, the arbitrator might prudently reserve that function (if not agreed) as well as quantum, in which case the following might replace paragraph 50:

50   I RESERVE to my further or final award (if not agreed):

    (a) the allocation of costs under paragraph 48(d)(i)–(iii) above; *and*

    (b) determination of the amount(s) of recoverable costs; *and*

    (c) any interest thereon;

for which purpose I shall give my further instructions upon the application of either party.

6.4.3   *Counterclaim variants*

There can arise many variants on the scenarios just illustrated, but part of one should suffice for the purpose of illustration.

---

5   See provision re interest at 6.3.2 illustration and footnote caveat thereto.

Scenario

Imagine that the counterclaims in 6.4.1 and 6.4.2 had *both* applied, but that which related to damage to the wall had instead been a claim for the estimated cost of repair and that, for present purposes, had been assessed at the time of the award, thus not being subject to the addition of interest. The amount awarded was £4,800.00. The award on the counterclaim (set-off) for the redecoration remains unaltered at £1,100.00, with £29.29 interest.

The redecoration, being a set-off, is still amalgamated with the claimant's claim as a single event, but the award relating to the wall, being a true counterclaim, is a separate event.

The summary could therefore be, for example:

Summary

43/ The amalgamation of these figures gives:

The Claimant's claim:		
Awarded		£17,100.00
Interest		269.38
		£17,369.38
Less: Set-off	£1,100.00	
Interest	29.29	1,129.29
Amount to Claimant (claim less set-off):		£16,240.09
The Respondent's counterclaim:		
Awarded		£4,800.00
Interest (not applicable)		n/a
Amount to Respondent		£4,800.00

If the operative paragraph 48 were to follow the form of 6.4.1, but avoiding the complications which would arise if details were incorporated, 48(a) could be:

(a) the Respondent shall...pay to the Claimant the sum of £11,440.09 (eleven thousand four hundred and forty pounds and nine pence), being the amount awarded under the claim less that awarded under the counterclaim; *and*

Introduction to further illustrations

As indicated previously, the illustrations which follow in this chapter (and in Chapters 7 and 8) are restricted in the main to the operative sections and such elements as are essential for the understanding of those brief indications.

The intellectual process outlined in Chapter 4, i.e. that of considering submissions, of weighing evidence, and of ascertaining the facts and the relevant law to the extent that each is applicable, is equally relevant when considering the non-monetary awards discussed below and in later chapters. The illustrations which follow cover only the substantive element of the operative section. The remaining elements are discussed at 5.3.11. As to costs in these types of awards, see 4.3.4.1(2).

6.5 A declaratory award

Examples of matters dealt with by way of a declaratory award are shown at 2.2.2. The general principles outlined in Chapters 3–5 apply, appropriately adapted – but in this form of award no money is ordered to be paid, and no 'instructions' are given. A declaration is made. Such a declaration might well have financial implications and commonly will do so, whether by subsequent action by the parties themselves or by further reference to the arbitrator. If the consequence of such an award is that an amount of money will become payable, the matter of interest should also be considered (section 49(5)). The consideration of that matter should follow the pattern for a monetary award, but such interest would be awarded by a further declaration, stating how such interest was to be computed.

In some instances, one party will seek a declaration and the other will simply oppose it. In other instances each party will seek different declarations. In (and in preparation for) the award, the background, the contentions, the arbitrator's analysis, and so on, can and commonly will take a similar form to that described in the relevant sections of preceding chapters. Section H only is illustrated here.

Scenario 1
STL claim that a term which had been discussed during pre-contract negotiations was, by reference, incorporated in the contract conditions for the mobile sets. That term was that TGP would without additional charge provide the services of a technician for one working day per month for three months to help train STL staff in the use of the sets. TGP simply reject the allegation.

In those circumstances, the claimant's claim either succeeds (or succeeds but in different terms) or it fails.

If it succeeds, the equivalent of the substantive part of paragraph 48 in the preceding illustrations might be:

48   I AWARD AND DECLARE THAT the contract incorporated a term that the Claimant (TGP) would provide the services referred to in its Statement of Claim at paragraph [*number*];

If it fails, and here assuming that it was one of several issues, the relevant part of paragraph 48 could be:

> 48 I AWARD AND DECLARE THAT the contract did not incorporate a term . . .

or in some circumstances it could be appropriate simply to award and declare that the respondent's claim (identified) fails.

Scenario 2
The same as Scenario 1 but STL, rather than a blank denial, allege that provision of service under such a term was conditional upon it being satisfied that there had been 'substantial malfunction' of a set, that being a condition precedent – and that there had been no such malfunction.

One of the possibilities could be:

> 48 I AWARD AND DECLARE THAT:
> (a) the contract incorporated a term that the Claimant (TGP) would provide the services referred to in the Respondent's [Statement of Defence] at paragraph [*number*];
> (b) such provision was conditional upon proper notice of material malfunction of one or more set(s); *and*
> (c) there had been no such malfunction; and in consequence
> (d) the said term is inapplicable.

6.6 A performance award

A brief explanation of a performance award, and a caveat, is set out in Chapter 2 at 2.2.3.

Scenario
STL contends that TGP undertook under the contract to provide an operator's manual explaining the computer software which controlled certain aspects of the operation of the sets and that TGP had failed to supply such a manual. STL seeks a performance award requiring TGP to remedy the deficiency.

This illustration assumes that the arbitrator had ensured (preferably as soon as possible after this issue arose in the arbitration) that, in the event of alleged subsequent failure to comply with such an award if issued, he was empowered to order appropriate further submissions and, if he found it appropriate, issue a monetary award to resolve the matter.

If the application for a performance award was successful, the equivalent, or relevant part, of paragraph 48 might be (the first part of the illustration applying only if the existence of such a provision was denied):

48 (a) I AWARD AND DECLARE THAT the contract incorporated a term that the Claimant (TGP) would on commissioning the sets provide the operator's manual referred to in the Respondent's Statement of Defence at paragraph [*number*]; *and*

(b) I AWARD AND DIRECT THAT the Claimant shall [*by when*] deliver such manual to the Respondent;

and in the event that there is alleged failure so to do or to meet the contractual requirements therefor, I reserve to my future order(s) and Award such damages as might be appropriate in the circumstances;

6.7 An injunctive award

An injunctive award is referred to briefly at 2.2.4.

Scenario
TGP had temporarily retained a mobile workshop in the theatre car park. The contract terms made provision for that facility and included a requirement that 'neither the workshop nor its use should be allowed to interfere in any way whatsoever with the use and operation of the theatre' (the 'non-interference condition'). The workshop was equipped with a radio link to TGP's development department; the radio was known to be faulty. The mobile sets housed equipment which relayed signals from actors' radio microphones. On several occasions during rehearsals the theatre sound system had picked up transmissions from the workshop which were audible over actors' voices. STL sought an injunctive award requiring TGP to take such steps as were necessary to prevent its radio equipment from causing interference to that of STL. TGP admits the existence of the contract condition.

If the application for an injunctive award was successful, the equivalent, or relevant part, of paragraph 48 might be (again with the safeguard in event of non-compliance):

48 I AWARD AND DIRECT THAT the Claimant (TGP) shall forthwith refrain from acting in breach of the admitted 'non-interference condition', specifically that it shall take all necessary steps to prevent radio interference caused by its equipment to that of the Respondent;

> and in the event that there is alleged failure so to do or to meet the contractual requirements therefor, I reserve to my future order(s) and Award such damages as might be appropriate in the circumstances;

6.8 A rectificative award

This type of award (rectification, setting aside, or cancellation) is described at 2.2.5. The first of those is illustrated here; the others could follow a similar approach.

Scenario
The accepted quotation had been made up of several items, all priced in £ sterling. The summary was arithmetically correct, but a TGP typist had erroneously hit the wrong key, thus showing the $ sign. After a petty, and unconnected, argument between the parties' respective accountants, STL's accountant had, mischievously or not, threatened to pay the final sum on the basis of changing the pound sign on the invoice to dollars. The relationship worsened and due to intransigence on both sides, the matter became an issue in the arbitration, TGP seeking rectification of the 'offending' document.

Again if successful, the relevant part of paragraph 48 might be:

> 48   I AWARD AND DIRECT THAT the document in question, being the summary page of Document [*number*] be rectified by correcting the $ sign in the total to £, leaving the figures as they appear;

7 Supportive awards

Awards on jurisdiction – Agreed awards – Awards on separate issues – Awards on reserved matters – Corrective or additional awards – Awards following remission – Awards giving further reasons – Awards following payment during the reference – Awards following an earlier adjudication – Unreasoned awards

7.1 Introduction

Comments and illustrations relate to English law and, unless stated otherwise, are on the basis that all 'default' provisions of the 1996 Act apply without alteration and that there are only two parties.

The illustrations in Chapters 6–9 relate in part to the same basic scenario as is set out at 5.2, but adapted so as to provide an extended or amended background for each succeeding illustration. See at 5.1, 'Introduction to the illustrations', for general explanations.

The further illustrations in this and later chapters concentrate on the impact of the changed scenario upon the operative section of each type of award, with a minimum of narrative explanation or illustration of other related award content. Incidental consequential changes to the underlying content should be self-evident to the reader, and it would be unnecessarily space-consuming to set out each and every change. Whilst in practice there might well be further interlocutory matters to recite as a consequence of each change to the scenario, such changes are ignored for present purposes (but see Chapter 9).

The various illustrations should be read in conjunction with Chapter 2, and with Chapter 5 where some of the non-illustrated narrative comments in that synthesis will come into play or have relevance to the illustration. Cover and title pages are not set out here. These would follow the appropriate general content illustrated previously, but incorporating the relevant type of award heading such as those illustrated, e.g. 'Agreed Award', 'Additional Award', 'Correction to Final Award', etc.

As matters other than the party relationship, form of appointment, etc. illustrated in Chapters 7–9 tend to diverge more from the underlying scenario, the suggested location of each illustration is indicated by the relevant checklist reference, thus:

[B1(j)]

[*no.*]   The Respondent challenged . . .

The further awards illustrated would still need to include or to reserve the decision on costs and any interest thereon (see for example 6.4.2.6, and also see 7.5).

The intellectual process outlined in Chapter 4, i.e. that of considering submissions, of weighing evidence, and of ascertaining the facts and the relevant law to the extent that each is applicable, is equally relevant when considering the non-monetary awards discussed below and in later chapters. As to costs in these types of awards, see 4.3.4.1(2).

7.2   An award on jurisdiction

Examples of matters dealt with by way of an award on jurisdiction are shown at 2.3.1. The general principles outlined in Chapters 3–5 apply, appropriately adapted, but in this form of award no money is ordered to be paid, and no 'instructions' are given. The award is in effect a declaratory award or declaratory part of an award, giving a ruling. Such an award can be dealt with either (a) in an award dealing solely with jurisdiction, at the time of or shortly after the challenge, or (b) as part of a subsequent award dealing with the substantive issue(s).

That part dealing with the issue in question, jurisdiction, would follow a substantially identical route in both forms of award – setting out the contentions, the facts, the law and its application to the facts, then a conclusion. If the award deals solely with jurisdiction, the 'background' would be recited only in sufficient detail as to identify the parties, the 'appointment', and sufficient of the procedures and matters in issue as to put the challenge into context. If jurisdiction is dealt with in an award on the substantive issues, the position of the jurisdictional issue in the award could depend upon when the challenge arose, but it should be dealt with at a time and in a manner which will avoid waste of costs in the event that the arbitrator's decision is that he does not have jurisdiction on the specific issue (or on the entire dispute if that is the extent of the challenge).

The illustration assumes that the objecting party has not seen fit to withdraw its objection and to agree, on grounds of economy, to an expansion of the matters referred to this arbitration. A simple situation has been adopted intentionally for this illustration, but such challenges can be much more complex, needing substantial submissions and evidence, and detailed analysis.

Scenario
The joint invitation from TGP and STL to Jonathan Fairley to act as arbitrator incorporated a brief statement of the matter in issue, i.e. that dealing solely with the matter of payment for redecoration works. The arbitrator accepted that appointment. When the claimant's statement of case was received it included an item relating to alleged disruption by the respondent of the progress of the installation and commissioning of the sets. The respondent's solicitor objected, contending that the matter was outside the clearly defined issue which was the subject of the arbitration – and furthermore that whilst an allegation of alleged disruption had been made at the relevant time, it had not been followed up by the claimant and there was no dispute as such on that matter at the time of the arbitrator's appointment. The arbitrator was asked (by both parties) to deal with the matter of jurisdiction as a preliminary issue.

In those circumstances the facts and the law would clearly lead the arbitrator to decide that the disruption issue was outside his jurisdiction. His award on that issue could be quite short, reciting only sufficient background as to identify the parties, the underlying dispute, the matters which led to the challenge, the submissions on the preliminary issue, then his analysis and decision. The operative section H might be (although in view of the wording of section 30(1) and 31(4) an arbitrator might prefer to use the words I RULE THAT . . .):

[H(b)2]

[no.]   I AWARD AND DECLARE THAT the said disruption issue does not form part of this reference to arbitration and is beyond my jurisdiction.

It could either go on to deal with costs of this award or could reserve them to the award on the substantive issue(s).

If this matter of jurisdiction were to be left over to the (or a) substantive award, this issue could to advantage be incorporated in its entirety at an early stage (B1(j)).

7.3   **An agreed award**

The provision for an 'agreed award' following a settlement is briefly described at 2.3.2.

It is obviously necessary to identify the parties, their relationship, the dispute and the appointment. That might be substantially as Sections A and B1 of the Chapter 5 illustration.

Unless agreed otherwise by the parties, the general provisions and requirements of the 1996 Act relating to awards apply equally to an agreed award, in addition to which (by section 51(2) and (3)) its content must be agreed by the parties and it must state that it is, nonetheless, the arbitrator's award. The arbitrator does have the facility, however, (section 51(2)) of objecting to such an award being published in his name. This is not the place to expand at length upon such a situation, other than to say that commonsense should normally prevail and enable a mutually acceptable draft to be agreed. The two principal ways in which the draft of an agreed award might be generated are referred to in 2.3.2.

In the first of those two situations, where a substantially complete draft is provided to him, if the arbitrator considers that such draft fails to satisfy the requirements of a valid, enforceable award, he must be mindful of his duties under section 33 and seek to have such shortcoming rectified by party agreement. In doing so he must not by inept intervention by him risk any fracture of what might have been a carefully contrived settlement.

In the second of those situations, he must likewise seek to avoid any disruption to that settlement which might occur if he attempts to make *unnecessary* tidying adaptations of agreed basic wording.

Assuming, then, that the draft as such complies with all the necessary requirements, he should ensure that:

(a)   by the heading, or part of the recitals, it is made clear that this is an agreed award *unless* the parties have valid and legal reasons for it not being so described; *and*
(b)   by suitable words it is stated to be *his* award.

The first of those, subject to the 'unless' qualification, can substantially be met by the title, on the cover as well as the title page, describing it as an agreed award. He should then ensure that reference to having considered the evidence etc. (H(a), paragraph 47 in the Chapter 5 synthesis) is excluded and is replaced, under the 'agreed' heading, by something like:

Agreed Award

[H(a)]

[no.]   This is my award in accordance with section 51 of the Arbitration Act 1996, {recording a settlement by way of an agreed award}.

An agreed award would not normally give reasons (subject to the parties' rights under section 52(1)).

The operative section H as such would follow a similar pattern to that previously illustrated.

It is necessary to bear in mind, however, that the settlement might have left liability for costs to be determined by the arbitrator (section 51(5)), in which case the replacement illustration of paragraph 47 referred to above would need to have words added to the effect of:

> , except as to the matter of recoverable costs, which the parties have left to my discretion. My decision on those costs, together with reasons therefor, follows the substantive award below.

7.4   Awards on separate issues

This topic is discussed at 2.3.3.

7.4.1   *The first award*

The first award in such a series would, subject to the nature of the issue(s) addressed, be similar to those discussed previously, with 'full' recitals of relevant background. There would be an indication in the title and in the text that this was the first of a series of awards, specifying which issue(s) are dealt with by it. In the interlocutory procedural matters (B2(g)) or other logical place, there could be, for example:

> **FIRST AWARD**
> **(relating to Issue No 1)**
>
> [B2(g)]
>
> [no.]   This award relates to Issue No 1 [*description*] only.

7.4.2   *Subsequent awards*

Subsequent awards in the series would similarly indicate the location in the series and the issue(s) dealt with. The award which determines the last remaining substantive issue(s) would be headed 'FINAL AWARD' (or 'FINAL AWARD, except as to costs', or 'FINAL AWARD, except as to quantum of costs'), whichever is appropriate. In those subsequent awards there would generally be no need to repeat the whole of the background. That could be done by reference, e.g:

> [B(+)]
>
> [no.]   The contract between the parties, the dispute, my appointment and other relevant background were recited in my First Award dated [*date*]. That award determined Issue(s) [*number(s)*], reserving all remaining issues, and costs, to my further award(s).

However, in the intervening period matters could have arisen which *should* be recited in this award. These might relate to changed procedures, withdrawal or settlement of one or more issues, details relating to the issues dealt with in this award, default subsequent to the first award, and so on.

Otherwise, the award could continue with:

[B(+)]

[*no.*]   This award determines those remaining issues, viz Issues [*numbers*].

and, if appropriate:

, except as to costs.

The operative section H would again follow a similar pattern to those previously illustrated.

7.5   An award on reserved matters

This topic is discussed at 2.3.4.

The comments relating to awards on separate issues (2.3.3) apply equally here.

For present purposes, these awards are divided into: (7.5.1) awards on reserved matters other than the amount of recoverable costs; and (7.5.2) awards on the reserved determination of the amount of recoverable costs.

7.5.1   *Award on matters other than the amount of recoverable costs*

There is little need to expand upon the general run of awards on reserved matters. As indicated, they are much as described in 2.3.3 where relating to subsequent awards in a series. Additionally, the fact of some decision or determination having been reserved from a stated award (Section I), and that the reserved matter is addressed in this present award, would be recited (see for example the illustrations at 7.5.2 below).

7.5.2   *Award on the reserved determination of the amount of recoverable costs*

If this award concludes the reference it would be appropriately headed, such as:

FINAL AWARD

Determination of the amount of recoverable costs

However, in a long-running case of substance it is possible that certain costs, and determination of their amount, can arise on one or more earlier occasions. In such a case the heading for such earlier award might be, for example: 'THIRD AWARD Determination of recoverable costs arising and payable under Order No [*state number*]'.

Reverting to any award which deals with the reserved determination of recoverable costs, what needs to be recited is much as described at 2.3.4. The fact of having reserved this function would also be recited:

[B2(f)]

[*no.*] My [*state which and date*] award, having determined liability for recoverable costs, reserved to my further or final award the determination of the amount of such costs failing agreement between the parties thereon.

The party due to be reimbursed part or all of its recoverable costs would have prepared and submitted to the other party its appropriately detailed bill of costs. The 'paying' party might have objected to certain items or the amounts relating to them. If attempts to resolve their differences had failed, either (or both) of them could have applied to the arbitrator to act under that reserved function:

[B2(g)]

[*no.*] I have been notified that the parties have failed to agree on the amount of the [*state which party's*] recoverable costs and have received an application from [*state from which party, or both*] for me to determine that amount.

[B2(i);(j)]

[*no.*] I ordered and have received submissions from the [*objecting party*] specifying the items in respect of which objection is made, with its objections and grounds, and a response from the other party.

continuing with a brief outline of the process (on documents, or with oral addresses by advocates).

What then follows is in essence similar to a *reasoned* award, unless otherwise agreed by the parties. The arbitrator determines each of the issues raised. The manner in which he sets down his determination will depend on the outcome:

(a)   he might find no fault with the bill, in which case he would so state and determine at that figure;

(b)   he might find that reductions should be made to a relatively small number of items, in which case he would be likely to summarise his determination by deducting those reductions from the bill as submitted and determine at that reduced figure;

(c) he might make reductions to a majority of the items, in which case he might recast the whole as amended, but this can be complex; *or*

(d) (sometimes to be preferred) he might make all the decisions *in principle* (either, if there were a hearing, as the matter proceeds, or otherwise by an initial award on the matter), requiring the parties to complete the resultant calculations and adjusted amount, and to:

    (i) agree the resultant amount and simply confirm the fact of settlement but without details, in which case he could formally close the matter (see 8.2.2.1 and .2 suitably adapted); or

    (ii) notify him of the agreed amount for incorporation in a formal monetary, or declaratory, award on the amount of costs (as an agreed award); or

    (iii) refer any remaining differences to him for determination and final award.

As to interest on such costs (1996 Act section 49(4)), the underlying award should have included a declaration such as that illustrated at 5.3.11.7. In consequence, the decision would have already been made in principle by the related paragraph 48(d) and specifically reserved in paragraph 50 (see 5.3.12). The calculation of interest would form part of and be incorporated in this present award unless, in complex circumstances, it in turn was reserved for decision after further submissions.

Having appropriately briefly recited the process which had been followed, the operative section (H(g)) might be:

> [H(g)]
>
> [no.] I AWARD AND DETERMINE THAT, having by my [*state which*] award ordered that the [*state which party*] shall pay the [*state which party's*] recoverable costs and any interest thereon, the said recoverable costs amount to £[*amount, figures and words*] and the said interest thereon to the date of this award is £[*amount, figures and words*], thus totalling £[*amount, figures and words*], which sum shall now be paid by the [*party*] within [*number*] days from the date of this award.

Then, regarding post-award interest, if that is to be awarded:

> [no.] These amounts shall be subject to the addition of [*state simple or compound*] interest at [*state rate and with what rests*] from the date of this award up to the date of payment.

The award would then (a) deal with the arbitrator's remaining fees and expenses and (b) be formalised in the normal way and, if appropriate (i.e. if now a final award), conclude with:

[*no.*]   This my Final Award made and published by me in [*place*] on [*date*].

7.6   A corrective or additional award

This topic is discussed at 2.3.5. Before exercising the power to make either of these types of award, the arbitrator must give the parties (not solely an applicant) a reasonable opportunity to 'make representations' (section 57(3)).

7.6.1   *A corrective award*

'Award' in these circumstances can be a misnomer, for the provision in section 57(3)(a) of the 1996 Act is not for a new award as such, but a power to '*correct an award so as to remove . . .*'. However, by section 57(7) any such correction shall form part of the award.

Such correction is commonly likely to be of a much more constrained effect than an additional award. Provision for correction is restricted to the situation set out in section 57(3)(a). Whilst the consequence can be profound, the most likely situations are where there has been a simple accidental transposition of words, e.g. 'Respondent' where the context clearly indicates that it should read 'Claimant', or, having carefully reasoned through an issue and shown that the answer (possibly to a question posed in the negative) is clearly 'No', the award erroneously shows 'Yes'.

In view of section 57(7) it is preferable, if perhaps not essential, that the correction has the attributes of an award. That includes the giving of reasons if the correction is of a nature which warrants them. The heading could be, here assuming it relates to a final award:

Correction to Final Award dated [*date*]

The recitals need only be sufficient as to positively identify the parties, the particular dispute, and the specific award. This can be done by reference, as in previous illustrations of awards subsequent to a first award. That would be followed by a brief identification and explanation of the clerical mistake or error. In turn, that would be followed by any further necessary explanation and any reasons which would have been included originally had the error not occurred. The provision cannot be used, however, to give effect to reconsideration of the decision as such.

What amounts to the 'operative' section might in one of the simple situations indicated above be:

[H]

[*no.*]   In view of the foregoing, the word 'Respondent' in the first line of paragraph [*number*] in my Final Award dated [*date*] shall be corrected to read 'Claimant', and such correction shall now form part of that award.

A prudent arbitrator would wish to ensure that the previously issued erroneous award cannot be misused, unintentionally or not. The only reasonably positive way would be to secure the return of the faulty award and issue a corrected award, stated therein as constituting a replacement.

7.6.2   *An additional award*

A dispute might be the subject of a series of awards on separate, or separate groups of, issues. Failure by the arbitrator to deal with an issue which had been presented to him for inclusion in an award other than the final award could be simply remedied by the publication of a further award in the series. That would have a brief additional recital stating the fact of its previous omission. In other respects the recitals could be incorporated by reference as indicated at 7.4.2 above.

In other situations, or where what purported to be a *final* award but which missed that issue had been published, the subsequent additional award would need to be so entitled. Its nature and how it had arisen should be explained in the recitals, also making it clear (if that were the case) that this award is now the final award:

Additional Award
(relating to Issue [*number*] omitted from the Final Award dated [*date*])

The recitals need only be sufficient as to positively identify the parties, the particular dispute, and the award to which the additional one relates. That can be by reference, as in previous illustrations of awards subsequent to a first award. The explanation for the additional award, if one were felt necessary, would depend upon the circumstances, but might be, for example:

[B]

[*no.*]   By agreement between the parties, Issue [*number*] was taken out of sequence at the hearing relating to matters for my first award, but I was requested to defer its determination until a later award in the intended series. By mischance a determination of that issue was omitted from what

> purported to be my final award. This Additional Award rectifies that omission and, together with my last previous award, now constitutes my Final Award.

The structure of the award would then follow that of the appropriate type illustrated previously (monetary, declaratory, etc.), as would the operative section H.

If the omitted issue was monetary, it is possible that the incorporation now of the omitted issue could, in a claim and counterclaim situation, change the overall balance between claimant and respondent. The new summary (see for example summaries at 6.4.1.8 and 6.4.3) would show if that were so. It could mean that the balance of success changed so that a net payment became due in the opposite direction from that previously awarded. It could change liability for costs. That would be a serious situation and highlights the importance of preparing, and using and updating, a unique checklist and list of contents for each *separate* award within the arbitration and for the arbitration as a whole. The arbitrator would also need to check whether the 'mistaken' award had been acted upon by one or both parties and, after submissions, act accordingly.

If that issue was or related to, for instance, a declaratory component, it could usually be dealt with by a simple insertion directed to be made by the additional award. With any of the types of remedy, it would be imperative to invite submissions before rectifying the omission by issue of an additional award.

Whether the extreme situation described above applied or not, unless the necessary adjustment was of a simple nature it is likely to be best dealt with by stating in the additional award that the whole operative section, or a clearly identified part of it, is replaced by that set out in the additional award. The new operative section would be set out, incorporating the whole of the decisions replacing the excised part, having dealt with and shown the analysis and decision-making in the usual way. The overall possibilities are too variable for further specific illustration to be helpful.

In a simple situation the new operative section could simply but unambiguously set out an amendment.

The comment at the end of 7.6.1, relating to prudence, can also apply here, depending upon the circumstances.

7.7   An award following remission

This topic is discussed at 2.3.6.

A likely, but not the only, situation for remission is where the arbitrator has erred in law and the court remits the award or part of it 'for reconsideration in the light of the court's determination' (section 69(7)(c)). In making 'a fresh award in respect of the matters remitted' (section 71(3)), the arbitrator needs to decide whether that can be done as a correction, or by way of a completely new award replacing the existing one. See 7.6 above, but section 71(3) is specific – 'a fresh award'. Unless the matter remitted is a very small and self-contained part of a substantial award, and unless the required change in the remitted part cannot change other parts of the award (including costs), it is likely to be more cogent and certain, and would have full regard to section 71(3), if the whole award is replaced. That ensures that all is contained within a single document, that the corrected part sits comfortably with the remainder, that it takes proper account of any changed relationship between the claim(s) and any counterclaim, and that any change in liability for costs is properly dealt with.

The new award could have substantially the same content as that which it replaces *except that*:

(a) the heading would now be, for example:

> **AWARD replacing** [*state which*] **Award following remission**

and:

(b) an additional recital would be inserted, recording the publication of the now replaced award, the application to the court, the effect of the court's determination and the remission;

(c) a specific statement that this award follows reconsideration in the light of the courts' determination (in this illustration, on a point of law) and replaces that identified; that statement might be:

> [B1(l)]
>
> [*no.*] This is my amended award following reconsideration in the light of the said determination of the court and replaces my [*state which*] Award published on [*date*].

(d) the issue(s) affected by the court's determination and remitted by the court would be re-analysed (D) in the light of that determination, reaching a new decision; however, if proper regard to the court's determination nonetheless *for good reason* produced the same decision, that decision would be reiterated;

(e) the 'non-remitted' items would be repeated as before;

(f) the decisions on the various issues would be resummarised;

(g) any interest considerations would be readdressed;

(h) similarly with costs (having regard also to any decisions of the court on costs, including any on the arbitrator's fees);

(i)   a new operative section H would be prepared, merging any non-remitted items with the result of the reconsideration following remission, and adjusting accordingly.

7.8   An award giving further reasons

This topic is discussed at 2.3.7.

The arbitrator must have had reasons for his award, but if he has failed to show them, or is considered to have failed to show them in adequate detail, he can be ordered to state those reasons in sufficient detail for the purposes of the court – unless the parties have agreed to dispense with reasons. The requirement is possibly best met by an addendum to the existing award. A complete replacement award incorporating those reasons is feasible, but can generate other problems which are best avoided. It must be made clear that the reasons form part of the award and its original decision(s). This subsequent giving of reasons, or further reasons, cannot be treated as an excuse for a change of mind, but if complying with such task led to such a change of mind of serious consequence, the arbitrator would need to notify the court and seek directions – and face any consequences.

If the reasons are given by way of an addendum, that must be suitably identified, possibly:

> **Addendum to** [*state which*] **Award**
> **(Reasons for decision on Issue(s)** [*state which*] **)**

The underlying award should be sufficiently identified by brief recitals in a similar way to that described in 7.6 above and illustrated at 7.4.2.

The manner in which the addendum has arisen should then be set down, whether by order of the court or simply by request of the parties; if by the court, any additional elements of the court's order (e.g. relating to costs) should also be recited.

The reasons should then be set out, essentially in the same manner as would have been the case had they been included initially (see for example Chapter 5), but under a suitable heading or introduction. Here the issue from Chapter 5 is used as an example, but as though the reasons *had not* been given in that award:

> [D(c)]
>
> [*no.*]   I am ordered {I am requested} to give reasons for my decision on Issue [*number*], being by way of a declaratory award that the Claimant is entitled to additional payment for executing the work in dispute. My reasons for that decision are as follows:

> [Here would be set out the reasons, in the same way as were shown in Chapter 5, then concluding with:]
>
> [*no.*] The foregoing reasons now form part of my [*state which*] Award dated [*date*].

The Addendum would be formalised in the same way as shown in Chapter 5 Section J, but in place of 'This my First Award...' would appear 'This Addendum to my [*state which*] Award...'.

7.9 An award following payment during the reference

This topic is discussed very briefly at 2.3.8.

A respondent might during the currency of an arbitration make one or more payments to the claimant on account of one or more of the matters in issue. Circumstances can vary, but for present purposes it is assumed that the arbitrator is notified of that fact, by amendment of pleadings or otherwise, during or prior to the hearing. If for any reason the parties did not wish the arbitrator to be aware of those payments lest the knowledge might influence his decision, they might ask that he publishes a first award (determining, but not yet ordering payment of, any overall amount due) on all substantive matters other than, and reserving, interest and other matters to be the subject of further submissions, or the like.

Otherwise, the fact of the earlier payment(s) would be recited in the award (B3(h)) and the figure(s) so paid could appear as a deduction in the initial financial summary of the amounts determined (D(d)). In other respects the award would follow the pattern of a monetary award discussed earlier (Chapters 5 and 6).

7.9.1 *Where that payment made no provision for interest*

Scenario 1

As in Chapter 5, but assuming that:

- there had been several issues, mainly where liability was admitted but quantum contested;
- the amount claimed totalled £57,000.00, all of which was in dispute at the commencement of the arbitration, having been due for payment on 18 March [year];
- the respondent had of its own accord made a payment of £15,000.00 to the claimant on 4 August [same year];
- that payment made no provision for interest;
- during the hearing the arbitrator was made aware by both counsel of that payment and that the cheque had been cleared;

- the hearing had been later than in Chapter 5, i.e. was on 10 November [year], and the award was published on 28 November [same year];
- the arbitrator determined that, of the £57,000.00 claimed, his award should be £40,000.00 (Issue 1: £17,100.00; Issue 2: £12,400.00; Issue 3: £10,500.00); plus interest.

It is likely that pleadings would have been amended, reciting the payment and claiming only the balance.

In the award, having recited that background and the fact of the cleared payment, the summary (D(d)) could be:

[D(d)]

[*no.*] The amalgamation of these figures gives:

Issue No 1	£17,100.00
Issue No 2	12,400.00
Issue No 3	10,500.00
	£40,000.00
Less: paid 4 August [*year*]	15,000.00
Net amount awardable	
(before interest)	£25,000.00

Whilst, with the slightly longer period now involved, interest would be likely to be on a compound basis, that has been ignored for this illustration.

Then in Section F, Interest, having decided that interest (in this illustration adopting simple interest) should be added thus:

(a) under section 49(3)(a), i.e. on £25,000.00 awarded, from 18 March to 28 November; *and*

(b) under section 49(3)(b), i.e. on £15,000.00 paid before the award, from 18 March to 4 August;

and having set out the reasons (see Chapter 5 illustration in F, at paragraphs 41, 42), the decision on interest could follow:

[F(a)1–2;(b)]

[*no.*] In the exercise of my discretion, therefore, I shall award simple interest at [*rate*] per annum:

(i) on the sum of £25,000.00 from 18th March [*year*] until the date of this my award; *and*

(i) on the sum of £15,000.00 from 18th March [*year*] until 4th August [*year*];

and I compute those amounts at £873.29 and £285.62 respectively.

In that illustration the same rate of interest was adopted by the arbitrator in respect of both of the two elements (section 49(3)(a) and (b)). However, in times of volatile or substantial changes of underlying interest rates, he might feasibly have adopted different rates, reflecting such changes. That could be particularly so if the relevant periods spanned over a considerable time and/or the figures involved were substantial.

Another option in such circumstances would be to apply one rate on the whole £40,000 from 18 March to 4 August, and another rate on the balance of £25,000 from 4 August to the date of the award.

It might be helpful to incorporate a further summary after the illustration shown above, clearly to indicate the sum awarded:

[F(b)]

[no.]   The accumulation of the net sum awarded and all interest is therefore:

Net amount awardable	£25,000.00	
Interest:		
on £25,000.00		873.29
on £15,000.00		285.62
Total		£26,158.91

Costs would be as the illustration in Chapter 5. The operative section H (see Chapter 5 at award paragraph 48(a)) would be adapted accordingly.

By way of further example, if matters of VAT had not been relevant, the separate parts 48(a)(i) and (ii) (see 5.3.11.4) could be combined in a single item, also giving an inclusive total. In consequence, the operative part of the award (sub-paragraph 48(a)) in those circumstances could be:

48   I AWARD AND DIRECT THAT, in full and final settlement of all claims in this arbitration:

(a)   the Respondent shall within . . . pay to the Claimant the sum of £26,158.91 (in words), being the substantive sum awarded plus interest awarded, as summarised at paragraph [number] above; and

but if identification of any element which might have VAT implications was necessary, that could be done by cross-reference to the gross value figure (here £40,000) shown in paragraph D(d) above. See also at 6.4.1.10.

7.9.2 *Where that payment made provision for interest*

Scenario 2

All as Scenario 1, except that the payment on 4 August had been for £15,000.00 plus (and so designated by the respondent) £250.00 by way of interest.

The comment under the Scenario 1 discussion, i.e. regarding substantial shifts in underlying interest rates, can affect the manner in which the interest here paid on 4 August is treated. In any event, other than in the simplest of situations, it can more clearly demonstrate that such interest has been properly accounted for if it is shown separately in, e.g. award paragraph D(d) above and what follows it.

That paragraph might therefore become, for example:

[D(d)]	
[*no.*] The amalgamation of these figures gives:	
Issue No 1	£17,100.00
Issue No 2	12,400.00
Issue No 3	10,500.00
	£40,000.00
Less: paid 4 August [*year*]	15,000.00 + £250.00 interest
Net amount awardable	
(before interest)	£25,000.00

The decision on interest shown under Scenario 1 could be adapted so as to refer to the previous interest payment:

[F(a)1–2;(b)]
[*no.*] In the exercise of my discretion, therefore, I shall award simple interest at [*rate*] per annum:
(i) on the sum of £25,000.00 from 18th March [*year*] until the date of this my award; *and*
(i) on the sum of £15,000.00 from 18th March [*year*] until 4th August [*year*];
and I compute those respective amounts at £873.29 and £285.62 (subject to deduction of the £250.00 interest paid on 4th August [*year*]).

The further summary illustrated under Scenario 1 could then be, for example:

[F(b)]

[no.] The accumulation of the net sum awarded and all interest is therefore:

Net amount awardable		£25,000.00
Interest:		
on £25,000.00		873.29
on £15,000.00	285.62	
less paid	250.00	35.62
Total		£25,908.91

The remaining comments and illustrations under Scenario 1, suitably amended, would apply.

7.10 An award following: an earlier order giving provisional relief under section 39; or an earlier adjudication (e.g. HGCRA: see Chapter 2)

As indicated, both situations are discussed here only in respect of monetary disputes but, having regard to the types of relief available under such orders or adjudications, there can be wider implications. It is assumed in the illustrations and comments that the amount previously ordered to be paid has been paid and has been so confirmed.

These topics are discussed very briefly at 2.3.9.

7.10.1 An award following provisional relief under section 39

The arbitrator can, but only if so authorised by the parties, provisionally order 'any relief which (the arbitrator) would have power to grant in a final award'. That order must be 'subject to the (arbitrator's) final adjudication; and the (arbitrator's) final award shall take account...'. Note the dual function at the award stage – final adjudication and final award. The arbitrator must ultimately deal with the subject matter of that provisional order on the basis of such full submissions as the parties might wish to make. Normally, the issue(s) to which the earlier section 39 Order related would be readdressed as a whole, the decision being by way of an award finalising such issue(s).

Regarding a monetary order under section 39, see the note at the end of Chapter 8.

The format of an award following a section 39 order will be very similar to any other award, but should additionally recite (B2(l)) the fact of the section 39 application, the consequential order, the payment if the order was monetary or what had transpired if the order related to some other 'relief'[1], and sufficient of the circumstances as to put the

---

[1] See section 39(1) and comment at 5.3.4.12.

matter into context, going on to recite and consider the full submissions now made.

If the relief granted under section 39(1) had been non-monetary, the circumstances are likely to have been somewhat unusual. The subsequent award in that situation could need to effect complex unravelling. That is possibly a good reason for viewing the granting of non-monetary section 39 orders with considerable circumspection, but that is outside the purpose of this book.

An illustration of recital of the prior issue of a section 39 order appears at 9.3.4.8. That illustration relates to costs. If such an order had related to one or more substantive issues, the last phrase of that illustration would be amended to identify the relevant issue(s). In this present instance, it is assumed that Issues 1–3 are the whole of the issues. That illustration could then continue:

> [B2(l)]
>
> [no.]   Those issues are considered afresh below {the parties having delivered more
>   detailed submissions and having adduced additional evidence}.

After considering the full submissions and reaching a reasoned determination, described as 'adjudication' in section 39(3):

(a)   if his final adjudication[2] remains the same as his provisional order he should explain why, particularly if the submissions remain wholly or substantially unaltered from those in the section 39 application but he has now been asked to give further consideration to them;

(b)   if his final adjudication changes from that provisionally ordered he should similarly, in giving reasons based upon consideration of the more extensive submissions and evidence, explain that change.

A provisional order under section 39 might have related to one or more, but not all, issues in the arbitration. If the parties made no further submissions on the subject matter of that order, the arbitrator should check to ascertain whether they wished to leave such provisional determination(s) unchanged, i.e. broadly equivalent to an inter-party settlement on such issue(s). If they so confirmed, he would recite that, too, in his award and deal with the matter appropriately in summarising his decisions (e.g. see 7.9 above).

Whether or not a monetary section 39 provisional order had included an amount in respect to interest, the resultant payment

---

[2] Note that the term 'adjudication' as used in section 39 has a different meaning from that word when used in the Housing Grants, Construction and Regeneration Act.

would come within the provisions of section 49(3)(b). The arbitrator would have a discretion to award such interest and a duty to re-address the matter in accordance with the 'final adjudication; and the (tribunal's) final award shall take account...' requirements of section 39(3).

7.10.1.1 Section 39 order relating to substantive monetary claim
If the amount provisionally ordered had related to the substantive claim, the award could proceed much as described under 7.9 above, suitably adjusted if that provisional order had related (a) to part of the substantive sum claimed *and* (b) to interest on the amount so ordered. The arbitrator would obviously be aware of how much, if any, of the amount had been interest, so would be able to prepare the calculation accordingly.

Scenario
Assume for present purposes the figures and dates were, perhaps coincidentally, as in 7.9 above, but the £15,000.00 had been paid as a result of a section 39 order *and* that order had in addition provisionally ordered that £250.00 by way of interest be paid. The claimant's claim was for £57,000.00, amended after receipt of the section 39 order by a rider: 'This amount is subject to the deduction of £15,000.00 which, with £250.00 interest, was the subject of the provisional order under section 39, provided that such amount has been received by the Claimant, and has been cleared, prior to the arbitrator's award.' The claimant's counsel confirmed that the amount had been received on time and the cheque had been cleared.

That additional background would be recited in the award (B2(l)).

Having outlined the contentions, analysed the submissions and evidence, and reached determinations on the several issues (e.g. as illustrated in Chapter 5), the 'amalgamation' shown at (D(d)) might instead be:

[D(d)]		
[*no.*] The amalgamation of these figures gives:		
	Issue No 1	£17,100.00
	Issue No 2	12,400.00
	Issue No 3	10,500.00
		£40,000.00
	being	£25,000.00 awardable now
	and	£15,000.00 paid under section 39 order

Assuming that the arbitrator had computed interest as shown and described in 7.9 Scenario 1, the reasons would be set out, followed (here repeated for convenience) by:

[F(a)1–2;(b)]

[no.] In the exercise of my discretion, therefore, I shall award simple interest at [rate] per annum:

(i) on the sum of £25,000.00 from 18th March [year] until the date of this my award; and

(i) on the sum of £15,000.00 from 18th March [year] until 4th August [year];

and I compute those respective amounts at £873.29 and at £285.62 (subject to deduction of the £250.00 interest required under my section 39 provisional order and paid on 4th August [year]).

That could lead to a summary such as:

[F(b)]

[no.] The effect of my substantive decisions on these issues and interest thereon, adjusted for the impact of my Provisional order, is:

Substantive total		£40,000.00
Interest:		
under section 49(3)(a)		873.29
under section 49(3)(b)		285.62
		£41,158.91
Less, paid under section 39 Order:		
Substantive sum	15,000.00	
Interest	250.00	
		15,250.00
Now due, including interest		£25,908.91

Subject to the qualifications there indicated, the operative section H could be similar to that at award paragraph 48 in 7.9.1 above (but in the amount just summarised).

7.10.1.2   Section 39 order relating to costs

If the provisional order had related to costs (see section 39(2)(b) of the 1996 Act), in addition to reciting the background as described above, the operative part H(g) would be suitably adapted, possibly (here as an adjustment to that shown at 5.3.11.7):

[H(g)]

(d)   the Respondent shall pay the Claimant's recoverable costs, together with interest thereon at [*rate, simple or compound, period of rests*] from the date of this award until payment, the amount of such costs and interest thereon (in both instances after allowance for the amount(s) paid in compliance with my provisional order issued under section 39, and failing party agreement on the resultant amount), to be determined by me [*state basis*] by my further award, *and*

continuing as appropriate.

7.10.2   *An award following an earlier adjudication*

This topic is relevant to some construction industry arbitrations, but could also apply to any situations if a process exists whereby a decision can be made by an 'adjudicator' which is binding upon the parties unless and until one or other of them seeks to have it superseded by the court or an arbitrator.

In such a situation a subsequent award would have much in common with that described in 7.9 but, if the arbitrator so decided, specifically setting aside the adjudicator's decision and awarding such sum (or other claimed remedy covered by the adjudication) as he, the arbitrator, determined.

The manner in which the award is set out will depend to some extent upon just how the claimant has pleaded its case. Adopting only the figures from 7.9 above, and assuming that the three issues were the same in the adjudication and the subsequent arbitration, the claimant might in the arbitration claim the original £57,000.00 *less* £15,000.00 ordered to be paid (and since received) under the adjudication, the defence simply contesting that claim.

Alternatively, the claimant might claim the £57,000.00, knowing that the respondent was counterclaiming, by way of set-off, repayment of the £15,000.00 already paid. There could be various permutations, including a variety of possibilities relating to any interest included in the adjudicator's decision and any in the award. To some extent, these options are explored in the preceding illustrations. Any constraints on the award of costs will depend upon the rules of the applicable adjudication provision as affecting subsequent arbitration on the adjudicated issue(s).

There is no particular need for the award to have a special heading relating to the previous adjudication. However, the fact of the relevant scheme, the adjudication and its outcome should be recited, as should the manner of pleading in the arbitration (see preceding paragraphs), possibly in one or a combination of B1(f) or (g), B2(f) or (o), B3(h). It would also be recited, if it were pleaded that way, that the claimant had claimed to have the adjudicator's decision set aside and be re-placed by the arbitrator's award, considering the issues afresh and giving credit for the amount received under the adjudication.

The analysis could then substantially follow the format(s) shown in Chapter 5 adapted by 7.9 and 7.10.1 above. In view of the various possible permutations, no summary is illustrated here, but the two illustrations in 7.9 and 7.10.1 are indicative of possible methods of approach. If the adjudicator's decision was being set aside, the opera-tive section H would be likely to commence, after the introductory 'I AWARD...' and preceding the illustrative award paragraph 48(a):

> [H]
>
> [*no./letter*]   The Adjudicator's Decision identified at [*paragraph No*] is set aside.

and continue with the award which replaces it. In order to avoid confusion about the fate of the amount paid under the adjudication, the make-up of the sum now awarded to be paid (if there *is* a net amount due) could be shown or repeated in Section H:

> [H]
>
> [*no.*]   (a)   the Respondent shall within [*days*] pay to the Claimant the sum of £[*amount, figures and words*], comprising £[*amount*] determined under this my award, less £[*amount*] received by the Claimant under the Adjudication; *and*

It might, however, be that the arbitrator awards *in total* an amount to the claimant which is *less* than that decided by the preceding adjudi-cation (and which has been paid). That would be shown by inclusion of a summary similar to those illustrated earlier. In such a case section H would, after the 'setting aside' item, award a balance now due to the *respondent*, say:

> [H]
>
> [*no.*]   (a)   the Claimant shall within [*days*] pay to the Respondent:
>
> (i)   the sum of £[*amount, figures and words*], comprising £[*amount*] received by the Claimant under the Adjudication, less the total now determined under this my award, being £[*amount*]; *and*

In that case, of course, the consideration of and determination of liability for costs (G) would have had proper regard to that situation, including the fact that the claimant had continued to arbitration to no avail (a situation not unlike the rejection of an offer – see 9.3.10.1). The later part of the operative section H would reflect that decision on costs.

7.11 An unreasoned award

This topic is discussed very briefly at 2.3.10.

As there indicated, an award without reasons needs to recite at least (see Sections B1 and B2):

(a)   sufficient to identify the parties;
(b)   how the arbitrator came to have jurisdiction;
(c)   what was in dispute; *and*
(d)   the fact that the parties have agreed to an award without reasons.

Matters arising under Section B3 (Hearing) (see Chapters 2 and 5) need to be recited to the extent that they influence the validity of the award or any of its parts, but not otherwise. It should not be necessary (or appropriate) to recite more from Section C than sufficient under (a) (Contentions of the parties) to ensure clarity of the issues if they have not been adequately defined in the earlier parts.

The principal difference between a reasoned award and an unreasoned award is that neither Section D (analysis, etc.), nor the reasoning in F (Interest) and G (Costs), are shown in an unreasoned award, unless, and to the extent necessary to explain why, the award on costs does not obviously follow the event.

8
Procedural awards

Awards dismissing the claim – Awards on abandonment

Institutional awards

Awards under rules, or under other statutes – Awards under 'consumer' schemes

Ancillary awards

When is an award (perhaps) not an award?

8.1 Introduction

Comments and illustrations relate to English law and are on the basis that, unless stated otherwise, all 'default' provisions of the 1996 Act apply without alteration, and that there are only two parties.

The illustrations in Chapters 6–9 relate, where appropriate, to the same basic scenario as is set out at 5.2, but adapted so as to provide an extended or amended background for each succeeding illustration. See at 5.1, 'Introduction to the illustrations,' for general explanations. For explanations and qualifications relating to these further illustrations, also refer to the 'Introduction' at 7.1.

8.2 Procedural awards

8.2.1 *An award dismissing the claim*

This topic is discussed briefly at 2.4.1.

The situations in which this sanction can arise are possible in various permutations. The general comments set out below are, in common with the general purpose of this book, to highlight areas to be considered, rather than any attempt to set out the law.

Situations to which party default under section 41(3) or 41(6) of the 1996 Act could arise include those:

(a) where the default relates to the whole claim;
(b) where there have been a number of 'tranches' of hearing, each dealing with one or more issues in their entirety but with decisions left over for inclusion in a single award (perhaps then reserving costs) and the default arises only after one or more of those hearings; *and*
(c) where there has been a series of awards, each determining one or more issues, and the default arises after one or more of those awards have been published.

The sanction of dismissing the claim or counterclaim following such default is discretionary.

In the case of section 41(3) of the 1996 Act, any 'inordinate and inexcusable delay' in pursuing a claim, where that delay had the consequence set out in 41(3)(a) or (b)), could arise in any of those three situations. Unless other special circumstances also applied, it might appear to be unreasonable to apply that rigorous dismissal sanction to any earlier and separate parts of the claim in respect of which there had been no such default. In the absence of such special circumstances, the award dismissing the defaulting claimant's or counterclaimant's claim could apply to all those issues to which the default related, i.e:

(i) where (a) applies: all of the claims; *or*
(ii) where (b) applies: all those issues still to be heard to the extent that the delay related to them; *or*
(iii) where (c) applies: all the awards of those remaining to be published to the extent that the delay related to them.

In the case of section 41(6), 'failure to comply with a peremptory order...to provide security for costs of the arbitration' could again arise in any of those three situations. The default might occur at any time. Under situations (a) and (b) above, none of the issues would at the time of the default have been finally determined by an award of the arbitrator, so that dismissal of the defaulter's entire case might be appropriate. In situation (c), by section 58(1) an award (i.e. *any* award) is final and binding, so that it would appear impossible to dismiss the claim(s) determined by such existing award(s). However, if such earlier award(s) reserved costs, that matter of costs would be an issue still remaining and the defaulter's claim in that respect could be dismissed.

In common with any award, by section 52(4) an award dismissing a claim must give reasons, unless the parties have agreed otherwise.

8.2.1.1 An award dismissing the entire claim

If this was the end of the arbitration, such an award could be headed 'FINAL AWARD' or 'FINAL AWARD DISMISSING THE [*state which party's*] CLAIM'. If the non-defaulting party's claim(s) remained to be determined, then this would not be the end of the arbitration. In those circumstances, if it were a free-standing award, i.e. not combined with matters relating to the other party's continuing claim(s), it could be headed 'AWARD No [*number*] DISMISSING THE [*state which party's*] CLAIM', leaving the non-defaulting party's claim(s) to a later award.

As with any award, it is necessary to recite sufficient as to identify the parties, how they came to be in dispute and what that dispute was (B1 and such of B2 as is necessary to put the matter adequately into context). The award should then recite the circumstances which led to the default (B2(m),(n)), including any relevant orders and/or relevant circumstances, identify the default and recite any further orders or opportunities granted to the defaulter to explain or mitigate that default (B2(n),(o)). The other party's responses (C;D) and the arbitrator's consideration of such explanations should be briefly set out. Those recited facts would usually be sufficient reason for the award, but if there were further reasons they too should be shown.

Before proceeding to the operative section, that explanation which constitutes the reasons could conclude with a statement (the award which gives effect to that determination following at paragraph H below) to the effect of:

[D(c)]

[*no.*] **Against that background and in the exercise of my discretion, I determine that it is appropriate that I shall dismiss the** [*state which party's*] **claim.**

The arbitrator then has to consider the matter of costs of the arbitration. Whilst on the face of it the defaulting party would be expected to be liable for costs, there could be extenuating circumstances in view of which the arbitrator might exercise his discretion otherwise, at any rate to the extent of appropriate abatement. The reasoning and decision would be set down in the usual way.

That part of the operative section H dealing with the dismissal might be:

[H(b)2]

[*no.*] I AWARD AND DECLARE THAT the [*state which party's*] claim is dismissed.

That section would then continue with the award on costs or, if appropriate for circumstances which should be explained, a declaration that no award on costs is made, then the usual formalities (J).

8.2.1.2 An award dismissing certain (but not all) issues

If it were appropriate to publish a separate award dealing *only* with the issue(s) dismissed, such award could follow the pattern just described, but the recitals/reasons, determination and the operative section would all specifically identify the issue(s) to which it related.

If instead it were more convenient to amalgamate the dismissal of the identified issue(s) with an award on the remaining issues of claim or counterclaim, the recitals relating to the default etc. (as above) could be integrated into the other recitals. If that were the case, those recitals relating to default could be under a separate sub-heading with a cross-reference, at that point and in any list of issues, direct to the analysis, Section D. It could there be dealt with as indicated above. In any summary, such issue(s) could then be endorsed 'Dismissed'. In the operative section H, it could either form a separate section or could be a sub-item, either following or preceding the other substantive issues. In this illustration it is assumed that this award concluded all issues, i.e. those dismissed and those decided by the arbitrator:

[H(b)2]

[*no.*]   I AWARD AND DECLARE THAT Issue(s) No(s) of the [*state which party's*] claim is[are] dismissed, *and*

[H(b)1]

[*no.*]   I AWARD AND DIRECT THAT, in full and final settlement of all remaining issues in this arbitration:

   (a) the [*state which party*] shall within . . .

8.2.2   An award on abandonment

This topic is discussed in moderate detail at 2.4.2. The illustrations shown below relate to the various situations there indicated, and are subject to the caveats there expressed.

As with any award, it is necessary to recite sufficient as to identify the parties, how they came to be in dispute and what that dispute was (B1 and such of B2 as was necessary to put the matter adequately into context). That would be followed by a brief description of how the need for this award arose and any agreement as to how it was to be dealt with (B1(l);B2(e)).

8.2.2.1   Settlement in unknown amount

If after an apparent lack of action by both parties the arbitrator ascertains that they have settled the dispute, are prepared to divulge the terms and to agree to them being included in an agreed award, the position is in essence that described for agreed awards at 2.3.2 and illustrated at 7.3.

If, however, they do not wish to divulge the terms of settlement, the arbitrator (after being notified as described at 2.3.2 but without the terms) could publish a similar award but also reciting that, whilst the parties have confirmed that they have reached a settlement, they are not prepared to divulge the details – see below.

As to costs, including his own fees, he would seek to obtain their agreement to:

(a) deal with the allocation in accordance with their agreement if they would disclose it; *or*

(b) the matter being left to his discretion (which in the circumstances would seem unlikely); *or*

(c) him stating that, by agreement, he makes no order as to costs.

As to his own fees, and subject to the above, he might, if all else fails, award that in the absence of divulged express agreement his fees and expenses be 'paid as to one half by each party unless they have agreed, or now agree, that they pay in some other proportion'. Had he taken the prudent course of always ensuring that he held fees by way of security deposit, he could take that latter course, return any surplus in equal shares and leave it to the parties to sort out any other proportion.

After the basic recitals and description of how the present award has arisen, as described previously, the operative section H might be (but also see under 8.2.2.2(2) below):

[H(b)2]

[*no.*] I AWARD AND DECLARE THAT:

    (a) the matters in dispute in this arbitration have been settled by mutual consent in terms which the parties have withheld; *and*

    (b) immediately following the publication of this award I become *functus officio*; *and*

That would then continue with the matter of costs discussed above (or costs could precede (b)).

8.2.2.2   Abandonment as such

(1)   A 'Closing Order'

Arbitrations are sometimes simply abandoned at a very early stage, perhaps the seeking of an institutional appointment having been a lever to encourage the other party to negotiate or settle.

(a)   Abandonment before acceptance of appointment

The arbitrator-elect might have notified the parties of his fees before his acceptance of appointment but, having had no response, has not accepted the appointment. He can after appropriate enquiries simply notify the parties that he deems the matter closed. No award is necessary, or appropriate. He has no authority. It is unlikely at that stage that he could have incurred costs of any magnitude; he is not yet the arbitrator.

(b) Abandonment after acceptance of appointment

The arbitrator might have accepted the appointment, or it is a type of appointment which is effective from the date of it being notified to the parties by the appointing authority, but the parties have not responded to the arbitrator even after adequate efforts on his part. He might feel the need (see 2.4.2), after due notice and having allowed time for a response, to issue a 'Closing Order' (something for which, however, there is no statutory provision):

CLOSING ORDER

Such an 'order' would briefly recite the 'appointment', the failed attempts to communicate with or obtain responses from the parties, the notice to the parties of intent to issue this order if there was no further response, and the continued lack of response. The equivalent of the operative part might be:

[no.] In consequence of the foregoing, I hereby close this arbitration.

(2) It could have been an agreed award, but...

There can, however, be situations where such an order is either requested, or assented to, by the parties. That can result in the following type of closing order (which speaks for itself without further narrative explanation):

[no.] I was appointed arbitrator in this dispute [*state by whom and, if appropriate, that the parties accepted terms of appointment*] and accepted that appointment on [*date*].

[no.] The parties provided security for my fees.

[no.] During the pre-hearing proceedings I was notified that the parties had settled their differences, but that an award reciting the terms of settlement was not required. In consequence I sought confirmation that:

— any amount agreed to be paid had in fact been paid and cleared – and it has been confirmed on behalf of both parties that the answer is yes;

— the consent terms do not incorporate any future action or restraint by either party – and it has similarly been confirmed that there is no future action nor restraint incorporated; and

> — the terms are in full and final settlement of all matters in dispute in this reference – and the answer has similarly been confirmed as yes;
>
> and it has been confirmed that any balance of security for my fees held by me, less my chargeable fees, is now to be returned to the parties in equal shares.
>
> [*no.*] In consequence of the foregoing I hereby close this arbitration.
>
> [*no.*] I determine my fees and expenses at £[*amount*] inclusive of Value Added Tax, and in accordance with the request from both parties the balance remaining from security deposited with me is herewith returned in equal shares to each party.

8.3 Institutional awards

8.3.1 *An award under rules* (or other statute)

This topic is introduced briefly at 2.5.1.

Arbitration rules usually relate to arbitrations in a particular field of commercial activity, be it maritime, construction, grain and feed trade, or some other area. They are designed to facilitate the conduct of such arbitrations. As a generality, they might cover matters such as commencement of the arbitration, nature of and timetable for submissions, nature of procedures, powers of the arbitrator, alternatives (e.g. on documents only, or on documents and inspection, or by way of full or restricted attended hearing), pre-agreement on any provisions of the 1996 Act which otherwise apply only if the parties so agree, any constraints on recoverable costs, form of the award, and so on.

The fact that an arbitration is under the provisions of Rules would usually be made clear in the title and in the early recitals (B1(f)), naming the Rules:

> In the matter of the Arbitration Act 1996
>
> and in the matter of an Arbitration
>
> under the Theatre Industry Contracts Tribunal Arbitration Rules
>
> between
>
> Techsolut (GB) plc   Claimant
>
> and
>
> Sophisticated Theatres Ltd   Respondent

The manner in which the rules came to be applicable, whether as a condition of the underlying contract or by subsequent agreement, should also be recited. Where the rules provide alternative procedures or the like, the option chosen by the parties should be recited, e.g:

> [B1(f)]
>
> [*no.*]   Whilst not directly applicable under the ad hoc contract between the parties, by subsequent agreement by exchange of letters they agreed that the Theatre Industry Contracts Tribunal Arbitration Rules should apply to any arbitration which might arise under the contract. That agreement applies to this arbitration. It was further agreed that alternative provisions numbers [*give references and topics*] thereof should apply.

The rules or the agreement incorporating them might have made provision, failing agreement by the parties on the name of an arbitrator, for appointment by the issuing body or another appointing authority. If the appointment had been by that method, that and the fact of the appointment would be recited. For illustration of such appointments, see 9.3.3.2.

If the rules make provision for what might otherwise be, or seem to be, unusual powers or procedures, and if any such unusual power is exercised, it and how it has been exercised should be mentioned (B2(d)).

Subject to the foregoing, an award under rules is much the same as the relevant type of award (monetary, declaratory, etc.) described in earlier chapters, subject to any specific constraints such as a limit on, or a procedure in place of, recoverable costs.

Similarly, if the arbitration is under some other statute which specifically provides for arbitration (perhaps in restricted types of dispute), that should similarly be shown, both in the title and by recital. The same comments apply as set out above with reference to rules.

8.3.2   *An award under a 'consumer' scheme*

The general nature of such schemes is briefly outlined at 2.5.2.

In general, the comments at 8.3.1 above apply but, in addition, the administration of the reference is usually by the appointing authority. That body is, in effect, a buffer between the parties and the arbitrator. It usually collects the submissions and documents from the parties, passes them in their entirety to the arbitrator and issues the arbitrator's award. It charges a registration fee for those services. Some such schemes are to be decided by the arbitrator solely on the basis of documentary submissions and evidence, sometimes together with an inspection. Some can provide for a consumer party to opt instead for an attended hearing, but if this happens any provision within the scheme to restrict recoverable costs is unlikely to apply (subject to that being so provided by the rules).

These arbitrations can be subject to a tight timetable, they are commonly monetary and relate to relatively modest amounts (subject to section 91 of the 1996 Act). The time for payment is often specified in the rules, and the schemes often have no provision for awarding recoverable costs as such but instead only the reimbursement (if appropriate) by one party to the other of any registration fee paid to the appointing authority administering the scheme. The arbitrator's fee is commonly paid by the administering body out of the fee it receives from the sponsoring trade body and, if so, is not dealt with in the award (although that fact may be mentioned). It is obviously essential for any arbitrator new to such a scheme to check its provisions carefully, for they vary. For illustration of references on documents only, or documents and an inspection of the subject matter (which are procedures not restricted to consumer schemes) see Chapter 9.

The reasons for any reimbursement (or not) of a registration fee need to be set out. They are the equivalent of such reasoning relating to recoverable costs in an orthodox award. If any part of the substantive amount awarded has already been paid by the respondent to the claimant, that must be accounted for in the award in a similar way to that indicated at 7.9.

Because of that background, awards under such schemes are often substantially shorter than those in commercial disputes, not least because far less needs to be recited. Nonetheless, the award must still have the attributes of a valid, enforceable award, with reasons unless the particular scheme provides otherwise.

The operative part of the award might thus (having earlier in the award reached a conclusion to that effect, and after any 'have considered the submissions . . .' introduction) be:

[H(b)1;(g)]

[*no.*]   I AWARD AND DIRECT THAT, in full and final settlement of all claims in this arbitration:

(a)   the Respondent shall pay to the Claimant £[*amount; figures and words*]; *and*

(b)   the Respondent shall also pay to the Claimant the sum of £[*amount; figures and words*] by way of reimbursement of the Claimant's Registration Fee paid in accordance with the Scheme.

8.4   Ancillary awards

8.4.1   *When is an award (perhaps) not an award?*

This topic is introduced at 2.6 partly for completeness and partly as a reminder or caveat. For the reasons there expressed, these topics (i.e. orders under sections 39 and 38(3)) are not further developed here. In both instances the 1996 Act refers in the text to *orders* but, in the case of section 39, with that historic side-heading referring to 'award'.

The distinction between an award and an order is not simply a matter of semantics, nor can an arbitrator cause an order to be an award, or vice-versa, simply by entitling it one or the other. The nature and status of the document is determined by what it purports with proper authority to do. For present purposes an order might be considered as an interlocutory device which is capable of being amended, or expanded, or of being converted into a peremptory order. An award must be final in regard to the matters which it purports to determine.

This book is not the place to expand upon the distinction, but for further clarification see the 1996 Act at sections 42 to 45 (Powers of the court in relation to arbitration proceedings) and sections 66 to 71 (Powers of the court in relation to an award).

As to section 39, it is difficult to see how a provisional and temporary arrangement can be the subject of an award, on which see the DAC (Departmental Advisory Committee on Arbitration Law) February 1996 Report at paragraph 202 and 203. Also see note below.

Regarding section 38(3) relating to security for costs of the arbitration, during the consultation process on the embryo 1996 Act, concern was expressed by some commentators that an arbitrator faced with such an application might in some instances become privy to knowledge of an offer of settlement with a risk that he could be, or might be thought to be, influenced by that knowledge. That fear was discounted by the DAC (February 1996 Report, paragraph 196).

The problem remains, however, that no matter how rigorously the arbitrator has put such knowledge from his mind, one or both parties might still believe that he had been influenced by that knowledge. Provided that both parties agree, the problem could be overcome by the parties, with the arbitrator's approval, appointing a separate 'costs security arbitrator' to deal with an application for such security.

Note
Whilst in this book I have avoided the quotation of specific common law authorities, leaving such matters to other textbooks, I feel that it would be helpful to suggest that any arbitrator faced with an application for payment on a provisional basis under section 39 (and having regard to section 81 of the 1996 Act referred to in Chapter 1 at 1.2) should at least be aware (directly relevant or not) of *Modern Trading Co Ltd* v. *Swale Building & Construction Ltd*, (1990) ADLRJ 174 and *SL Sethia Liners* v. *Naviagro Maritime Corporation (The Kostas Melas)* [1981] 1 Lloyds Rep 18 to which it refers. Whilst the cases dealt with pre-1996 Act 'summary' interim awards, an area where the wise were loath to tread, there are lessons still to be learned from the constraints laid down in those judgments.

Part E
Variants

9 Examples of other options and variations in content

9.1 Introduction

Comments and illustrations relate to English law and are on the basis that, unless stated otherwise, all 'default' provisions of the 1996 Act apply without alteration, and that there are only two parties.

In the illustrations in Chapters 5–8 the arbitration agreement and mode of appointment remained unchanged, with a few necessary exceptions which would be clear from the context. That was done so as to concentrate solely on the purpose of each type of award in turn. Other options and variations are explored via the illustrations in this chapter – contract or tort, different modes of appointment, and so on.

In the main, no scenarios are described. The use of section headings, checklist references and brief introductory notes to each illustration indicates the situations where the various options or variations might come into play.

These supplementary illustrations are not intended to be comprehensive, but simply indicative. They are not given for every item in the checklist, but give a flavour of what might be appropriate in a variety of circumstances. They should be read in conjunction with the notes in Chapter 5 and later chapters. Where an illustration impacts specifically upon, or generates the need for, a later paragraph, that is in the main illustrated immediately following the first entry.

9.2 Format of this chapter

For ease of comparison with illustrations in previous chapters, matters considered below are located under the most relevant checklist heading.

9.3   Options and variations

Section A

9.3.1   The heading   [A]

Whilst the somewhat stereotyped heading shown in Chapter 5 is in common use, there is no mandatory reason for any particular wording, provided that it gives the necessary information. It could instead be in simple language, e.g:

> **Arbitration between**
>
> **Techsolut (GB) plc**   **(Claimant)**
> **and**
>
> **Sophisticated Theatres Ltd**   **(Respondent)**
> **under the provisions of the Arbitration Act 1996**
> **FIRST AWARD**

That, if desired, could be followed by the arbitrator's name.

Section B

9.3.2   Background   [B]

This collective heading is shown again here simply as a reminder that the matters covered by Section B1–B3 might arise and/or be dealt with in various parts of the award depending upon circumstances. See comment at 3.3.2.

Section B1

9.3.3   Identification and jurisdiction   [B1]

9.3.3.1   **Identification of the parties/contract or other
            relationship**   [B1(a),(b)]
(1) A claim in tort/submission agreement   [B1(b)]
The earlier illustrations, with one partial exception, related to disputes in contract, but arbitrations can arise from an alleged tortious act or omission with no contractual connection, e.g. (unrelated to the Techsolut case):

> [B1(a), (b)]
>
> [no.]   The Claimant is [*name and address*] and the Respondent is [*name and address*]. The Claimant alleges that the Respondent has caused subsidence to one of the Claimant's substantial garden structures by causing or allowing

> serious water leakage from the Respondent's newly-constructed wildlife pond and that in consequence the Claimant has suffered expense by way of the cost of essential repairs.

and this might be followed by, for example:

> [B1(d),(e)]
>
> [*no.*] Having failed to reach agreement on either liability or the amount of damages flowing from the alleged tort, the parties have agreed by a written submission agreement dated [*date*] to appoint me, [*name etc*], as arbitrator in this dispute. Having agreed with the parties in writing my terms and conditions, I accepted the said appointment by my letter of [*date*].

This also embraces items B1(g) and (i). Details of what is claimed could be given here (e.g. see Chapter 5 at 5.3.3.8 (B1(h))).

9.3.3.2 Provision for appointment/appointment [B1(e),(i)]
(1) Appointment by an appointing authority [B1(e),(i)]

Where an arbitration agreement/clause provides for appointment by an 'appointing authority', the recital might be, for example:

> [B1(e), (i)]
>
> [*no.*] The said arbitration clause [*identify it if not already done*] provided that in the event of any dispute or difference (as therein defined) arising, such dispute or difference was to be referred to arbitration by a single arbitrator to be appointed, failing agreement, by the President or a Vice-President of the [*name the appropriate body*]. I was appointed arbitrator by the said President by a form of appointment dated [*date*] and accepted the appointment on [*date*].

In some instances such an appointment, or an appointment by the court, is effective on it being notified. In such an instance and if that is the case, instead of 'and accepted...' there might be 'that appointment becoming immediately effective'.

9.3.3.3 Basis for decision [B1(k)]

If, rather than as indicated in the initial note to 9.1, the chosen law was other than English law, a basic recital might be:

> [B1(k)]
>
> [*no.*] The parties have agreed that I am to determine the matters in issue in accordance with the substantive law of [*country*].

The award should then indicate the manner in which that law has been applied. That might be by indicating that the arbitrator is sufficiently knowledgeable in that particular law to act alone; he might have been appointed for that very reason Alternatively, for instance, he might recite that he has under the provisions of section 37 appointed a legal assessor, knowledgeable in that law, to assist him. The various possibilities and ramifications of other applicable laws are beyond the scope of this book.

If, alternatively, the parties have agreed in writing that the arbitrator's decision shall be on the basis of 'other considerations' (section 46(1)(b)), then depending upon what 'other considerations' have been agreed (here assuming one of the possibilities) an appropriate recital could be, for example:

> **[B1)k)]**
>
> [*no.*] The parties have agreed in writing that I am to decide Issues [*state which*] on the basis of justice and fairness having regard to established practice in the [*state what area of industry or commerce*] rather than in accordance with the law applicable to my decision on all other issues.
>
> [*no.*] In doing so in respect of those issues I invited the parties, in those instances where they differ as to the nature of that practice, to submit such evidence and authorities as they wish (other than legal authorities) as to the relevant practice.

See under Section D below for further illustration.

Section B2

9.3.4   *Interlocutory procedural matters* [B2]

9.3.4.1   Seat [B2(b)]
Two of the three options covered by section 3 are shown at 5.3.4.2, at paragraph 7 of the illustration. The other option might, as one possibility, be recited thus:

> **[B2(b)]**
>
> [*no.*] At the preliminary meeting the parties orally authorised me to designate the Seat of the arbitration. They had previously authorised me in writing under the provisions of section 5(4) of the Arbitration Act 1996 to record in writing any oral agreement relating to the reference and I duly recorded that agreement. Having heard such submissions on the matter of Seat as the parties wished to make, I hereby designate the Seat as [*state what juridical seat*].

9.3.4.2 Interlocutory agreement on rules, powers or procedures [B2(c)]

(1) On procedural and evidential matters [B2(c)1]

An example of one of the section 34 matters which can arise is, at 9.3.4.3 below, combined with an indication of how that power was exercised (B2(d)).

(2) On consolidation, or concurrent hearings [B2(c)2]

An example might be:

[B2(c)2]

[*no.*] The parties agreed that, as I had also been appointed arbitrator in another dispute ('the second arbitration') between them which had substantial elements in common with this arbitration, there would be concurrent hearings, but only on the issues which have those elements in common, being Issues [*state which*]. That second arbitration is the subject of one or more separate award(s).

Note that in this situation the agreement is for concurrent hearings, not consolidation of the two arbitrations. In this present case, then, there are still two separate arbitrations, even if with concurrent hearings. Unless agreed otherwise, there would be separate award(s) for each arbitration. Careful thought and planning would have been necessary in advance of the hearing. It would have been necessary to ensure that neither submissions nor evidence could have become blurred between the two cases.

In some instances, the issues, or sufficient of them, might be of an identical nature, with only the resultant computations differing. In other instances, the concurrent hearings might have been substantially a matter of convenience or economy in achieving necessary attendance (e.g. of witnesses) in a situation where the two cases had much in common.

Other than reciting the parties' agreement for, and the fact of, concurrent hearings, and how any procedural difficulties had been dealt with, the awards in the two cases could follow the usual pattern.

Had the party agreement been for consolidation of two or more arbitrations, there would be yet more scope for administrative complications. For instance, though not necessarily the case, party A might be claiming against B, who in turn alleges that C is liable not only for A's claim, but also a related claim by B. Consider the logistical problems if, in reverse, B and C have not just a defence, but also a counter-claim, in C's case partly against B and partly via B against A. Not only are pleadings and their sequencing complex in such a situation; so also is the preparation and drafting of the award or series of awards. The situation is mentioned here simply as a note of warning.

(3) On appointment of experts etc. [B2(c)3]

A situation relating to the scenario in the Chapter 6 illustration at 6.7 could have been that it was in issue as to whether or not TGP's radio link was capable of causing, or was in fact causing, the alleged interference. However, the arbitrator had neither radio nor electronics experience. The appointment *by the arbitrator* of an expert (section 37(1)(a)(i)) should be recorded, e.g:

[B2(c)3]

[*no.*] Having given the parties an opportunity to object to such proposal (at which time they concurred with it), I appointed Ms Dianne Farlington, an acknowledged expert in these matters, to consider and report to me and the parties on the linked matters of (a) whether the Claimant's radio link was technically capable of interfering with the audio function of the mobile sets, and (b) whether in fact, under test conditions, it did so.

[*no.*] Ms Farlington reported to me in the presence of the parties, and Counsel for the parties commented upon her findings. I adopted those findings and they form an element of the matters leading to my decision on the radio issue (see paragraph [*number*]).

(4) Relating to property... [B2(c)4]

An illustration, linked with that at (3) above, might be (and could conveniently be interjected between the two paragraphs):

[B2(c)4]

[*no.*] In order to facilitate that consideration, I ordered that such experiments as were reasonably necessary be carried out on the equipment in question (and the Claimant gave consent that such experiments should incorporate the use of its radio as such).

(5) On preservation of evidence [B2(c)5]

Linked with the two immediately previous illustrations could be:

[B2(c)5]

[*no.*] Mr Blackwelle, having told me that the stage crew maintained a timed diary of all problems with equipment and of disruptions to performances, expressed concern that the radio log apparently kept by the Claimant's site staff might, accidentally or otherwise, be disposed of before the two could be compared. I directed that the said radio log be preserved in a secure location on site until such time as I gave directions otherwise.

9.3.4.3 Use of 'uncommon' powers [B2(d)]

Whilst not necessarily an uncommon power, an example might be:

> [B2(d)]
>
> [no.] The parties agreed in writing that I should take the initiative in ascertaining the facts and the law in respect of such matters as in my discretion I considered to be appropriate. I opted so to do with regard to those issues which I was to decide on an equitable basis (being Issues [number(s)]). In so doing I notified the parties in advance of (a) questions which I intended to pose, and (b) information I should require, regarding the said established [trade] practice.

9.3.4.4 Meetings:...pre-hearing review [B2(h)]

In the illustration at 5.3.4.8 (B2(h)), brief reference to the procedural order(s) issued in consequence of the preliminary meeting were combined with the simple reference to the meeting having taken place.

Assuming that the arbitration was of sufficient magnitude or complexity to warrant a pre-hearing review meeting, an illustration could be (but this would often be better included at a later stage of the award):

> [B2(h); B2(i)]3]
>
> [no.] A pre-hearing review meeting was held on [date], attended by the respective Counsel and decision-making representatives of the parties. As a consequence of applications at that meeting following alleged default by the Claimant, I issued a peremptory order requiring the Claimant to comply by [date and time] with my Order No [number] relating to disclosure of a specified record of staff time expended on contract and extra works.

9.3.4.5 Consequent and subsequent applications and orders [B2(i)]

The Order to which reference has just been made could equally have been recited here. Similarly, the initial recital of such meetings as warranted mention could be restricted to the bare fact of such meetings having taken place. If necessary, that could also list who was present, leaving details of any orders or other action to be incorporated in the award at the most appropriate place for that particular award.

(1) Attended hearing or 'documents only'? [B2(i)1]

Depending upon the nature of the case, an illustration where there was to be no hearing might be:

> [B2(i)1]
>
> [*no.*] The parties agreed [*or requested?*] that this arbitration be conducted on the basis of written submissions, written witness statements, and documentary evidence.

If appropriate, this might be supplemented by (see 5.3.4.9(7)) for example:

> [*no.*] It was further agreed, and ordered by me, that:
>
> (a) witness statements be by way of sworn affidavit;
>
> (b) each party would have the facility to serve interrogatories; and
>
> (c) the answer(s), or any necessary expansion of a witness statement, be by way of affidavit.

If, as in this illustration, the matter was to be determined on documents only, the general description or categories of documents should be identified in the award, by listing or by reference (Section C). Any witness statements or reports should be specifically listed, identifying those which were by affidavit.

Such a list could be followed by a brief statement such as:

> [B3;C]
>
> [*no.*] Following close of written submissions I have considered the parties' written contentions and the documents tendered in support . . .

Then could follow an analysis and determination, generally similar to that illustrated in Chapter 5, but adapted to suit written submissions and evidence.

It might have been agreed that the matter was to be determined on the basis of, for example, written submissions together with an inspection of the subject matter by the arbitrator. That procedure is common in, but not limited to, some 'consumer' schemes. That agreement should be recited, in a similar manner to that set out above, but stating the function of the inspection. That could have been determined by party agreement, or by the rules of the scheme, if such it was. See 5.3.5.7.

The situation at 5.3.5.7 was that the arbitrator was simply, by that inspection, familiarising himself with the nature of the subject matter and its location. If, instead, the purpose was for him to use his specialist knowledge and/or expertise, such as in some commodity quality arbitrations, that, and the basis and extent of the arbitrator's relevant authority and inspection function, should be recited.

In making a decision on that basis (D), the arbitrator would need to describe what his inspection showed, his conclusion(s) and, if not self-evident from that description, his reasons for such conclusion. That conclusion, for instance, might be on whether the subject matter complied with the contractual requirements.

9.3.4.6   Resultant party action [B2(j)]
(1)   Result of meeting(s) of experts [B2(j)2]
A meeting of party-appointed experts of like discipline can have various results. A principal purpose of such a meeting is to ascertain (a) to what extent their respective opinions coincide, or would do so if specified facts were accepted, agreed or proved, and (b) to what extent they agree facts within their personal knowledge and expertise. An example of a consequent recital might be, for example:

> [B2(j)2]
>
> [*no.*]   The party-appointed experts [*names and by whom appointed*] on stage equipment met in accordance with my Order No [*number(s)*] and in consequence produced an Agreed Joint Report on Issue(s) No(s) [*numbers*]. They reported as their joint opinion that on Issue [*number*], if the facts relating to that issue were as contended by the Claimant, the failure of the equipment (the fact of which is not in contention) was due to a manufacturing fault, but if those facts were as contended by the Respondent, that failure could only have arisen from misuse by the Claimant's staff. They further agreed as a fact that the said failed component exhibited signs of hammer marks.
>
> As to Issue [*number*] . . .
>
> Their opinions differed as to all remaining issues, which remained as the relevant items in their respective reports.

9.3.4.7   Other directions and/or administrative action [B2(k)]
(1)   Orders for security [B2(k)1]
The fact of the issue of any order(s) for security for costs of the arbitration under section 38(3) can be recited here (B2(k)1), or later when dealing with costs. In any event, if there had been such order(s), the final disposition of that security needs to be dealt with in the operative section of the award. Such security can take a variety of forms, so the manner of dealing with that disposition will vary. Much has been written on the subject and specific study is warranted. There are so many permutations that to give illustration(s) here could be potentially misleading. However, see at 9.3.4.8 regarding a provisional order under section 39(2)(b) for interim payment (i.e. not simply security) on account of costs, and at 9.3.11.3 for release or disposal of security.

(2) Limitation of costs   [B2(k)2]

There should be no difficulty in reciting the simple fact (if that is so) of a direction that recoverable costs of the arbitration or any part of the proceedings be limited to a specified amount (B2(k)2).

Such direction must then be taken into account when later dealing with the determination of liability for recoverable costs (G), and in the operative section H when dealing with costs. Had there been such a direction in the illustration at 5.3.11.7, the subsequent paragraph 48(d) might have been:

> [H(g)]
>
> [*no.*] the Respondent shall pay the Claimant's recoverable costs up to but not beyond the limit on such recoverable costs imposed by my Order No [*number*], being a limit of £[*amount*], the amount of such recoverable costs and any interest thereon (failing agreement) to be determined by me [*state basis*] by my further award, *and*

9.3.4.8 Provisional orders   [B2(l)]

Recital of such a provisional order under section 39 relating to costs (here relating to a specific 'costs in any event' order, but if that criterion did not apply the last phrase of the illustration would not appear) might be:

> [B2(l)]
>
> [*no.*] Following application by the Claimant and acting under the authority vested in me in writing by the parties, and having heard submissions on behalf of each party, I ordered on [*date*], on a provisional basis under section 39 [of the Arbitration Act 1996], that the Respondent was to make to the Claimant an interim payment of £[*amount*] [on account of the sanction as to costs referred to in Order No(s) [*numbers*]].

Final disposition in the operative section H is illustrated at 7.10.1.

9.3.4.9 Party default/peremptory orders   [B2(m),(n)]

A recital relating to a peremptory order appears at 9.3.4.4 above, with another at 9.3.5.1 below.

Section B3

9.3.5 Hearing   [B3]

9.3.5.1 Failure to attend/ex-parte proceedings   [B3(c)]

Assuming that reference to an underlying peremptory order had been recited earlier, a recital at this point (here applying section 41(7)) might be:

[B3(c)]

[*no.*] Default by the Respondent (being failure to attend on the appointed day and time) and my consequent Peremptory Order have been recited earlier in this award. One of the consequences was that the Respondent was by that order put on notice of a new date and was warned that in the event that without good cause and adequate notice it again failed to present itself I should be minded to proceed in that party's absence. The Respondent having without notice failed to present itself at the said re-timed hearing, I delayed commencement for one hour and then, as warned, proceeded ex parte. The Respondent presented itself after the luncheon recess. As it failed to give any adequate explanation, I awarded the costs of and caused by its absence that morning to be paid by the Respondent in any event and proceeded with the hearing. Additionally, I warned the Respondent that the Peremptory Order was still in force and that if there was further failure in compliance without just cause I should proceed to an award on the basis of the entirety of the Claimant's case and only such of the Respondent's case as had already been supplied, drawing such adverse inferences from the Respondent's further default as the circumstances justified.

Section C

9.3.6 Submissions and evidence [C]

9.3.6.1 **Rules of evidence – relaxation** [C(c)2]
It is possible that the arbitrator had decided, or the parties had agreed, that the strict rules of evidence should not be applied. That might (if not a common event) have related to all evidence, or to that of a certain kind, or on a certain issue or topic, or from or on behalf of a particular party. That procedure would have been recited at B2(c)1 or other suitable place.

This kind of matter can, however, first arise mid-hearing, perhaps as a challenge to the admissibility or relevance of, or weight to be given to, specific evidence. Of various possibilities, that provision could result in, for example:

[C(c)2]

[*no.*] Ms Wellbie objected to the admissibility of certain evidence, being [*state what*]. I ruled that the evidence in question would be allowed in, but that I would give both Counsel a later opportunity to make submissions on its relevance and what weight I should place upon it.

Ms Wellbie indicated that she was satisfied that I had taken her point and was content that I would give that evidence only such weight as it warranted. Mr Blackwelle concurred.

The arbitrator would then in his analysis (D) state how he had treated that evidence.

Section D

9.3.7   Analysis etc.   [D]

9.3.7.1   Analysis under section 46(1)(b)   [D(c)]

Under the second illustration at 9.3.3.3, the arbitrator is to decide certain issues 'on the basis of justice and fairness having regard to established trade practice...'. In considering and analysing the submissions and evidence, and in reaching a decision, he must apply that same rigour as would apply to a decision based upon English (or other specified) law.

In finding facts, his function, and the manner of setting that down in an award, are much as illustrated under 5.3.7.3.

Having established and set down the relevant facts, instead of determining the relevant law, he must:

(a)   decide from the submissions and evidence whether or not there is an established trade practice relating to the matter in issue, and

(b)   in considering what is just and fair, incorporate and have proper regard to the effect of that trade practice (if such exists) in his reasoning.

To reach his decision on the relevant issue(s), he then applies to the facts the principles he has just determined.

Whilst in the illustration which follows, a trade practice of a practical nature has been relied on and adopted, it is likely that such trade practice in this context might relate to such matters as periods of credit, time for payment, responsibility for insurance, responsibility for obtaining any necessary statutory certification – a wide range of things. For this present situation, the appropriate award entry might be, for example:

[D(c)]

[*no.*]   On the basis of that evidence, I hold that there was, and is, an established trade practice in the stage equipment industry as to the sequence of assembly of components such as those in question. I have already found as a fact that the Claimant had followed that sequence.

That being so, it was neither just nor fair for the Respondent, during assembly, to seek to impose some sequence of its own choosing. I have already found as a fact that delay was occasioned by that intervention.

In consequence, I determine that the Respondent is liable for the delay occasioned by that intervention.

9.3.7.2 Appended reasons [C;D]

Relating to the positioning of reasons in an award, a specific variation from that illustrated in Chapters 5-8 is the situation where the reasons are set out as an appendix, or in some other manner at the end of the award. That might be as a result of requirements of the parties, or practice in a particular area of commercial activity, or any relevant rules, or the arbitrator's preference.

In such circumstances the basic award (i.e. other than the analysis which there demonstrates the reasons) would be very similar to an unreasoned award (but obviously not described as unreasoned), either in similar form to the remaining parts of that shown in Chapter 5 or in a condensed but adequately comprehensive form, with a statement in a suitable place to the effect that:

> [C;D]
>
> [*no.*]   My reasons are shown [appended] at [*state location*] and form part of this award.

Those appended reasons should be headed suitably to state that they form part of the identified award and should be signed. If the award proper was witnessed, the appended reasons should similarly be witnessed, preferably by the same witness, and if convenient should be bound with the award proper as a single document.

Section E

9.3.8   *Value added tax* [E]

9.3.8.1   VAT not applicable [E(a)-(c)]

If it is known that VAT is inapplicable, or if the parties have agreed that the arbitrator shall include no order relating to VAT, it can be prudent to insert in Section E, for example:

> [E(a)-(c)]
>
> [*no.*]   The parties have agreed that I make no order as to any Value Added Tax implications.

That could be followed, in the operative section H, by:

> [H(f)]
>
> [*no.*]   For the avoidance of doubt, the principal sum awarded at [*specify paragraph*] above is exclusive of any Value Added Tax.

9.3.8.2 VAT as a component of damages (awarded to a non-registered person) [E(a)–(c)]

If for example (see 4.3.2.1(2)) a non-registered person (for VAT purposes) is awarded damages for costs of repair and the charge for such repair is subject to VAT, that VAT component should be incorporated in the damages sum. The background could be recited in Section E:

> [E(a)–(c)]
>
> [*no.*] It was accepted on behalf of the Respondent that:
>
> (a) the Claimant is not registered for VAT;
>
> (b) in consequence cannot recover VAT charged to it on the remedial work awarded; and
>
> (c) the said VAT has been properly charged to the Claimant at the standard rate (17.5%).
>
> [*no.*] In consequence I shall add that VAT to the amount awarded under this item.

That VAT can be shown added into any summary of amounts awarded. In the operative section H that element of the award can either be shown at the VAT-inclusive figure and so described, or as £[*amount*] plus £[*amount*] Value Added Tax, thereby totalling £[*amount*].

If the cost of that repair, and its related VAT, had been paid by the non-VAT registered claimant pre-award, any interest awarded on the substantive sum would also apply to that VAT component.

Section F

9.3.9 Interest [F]

9.3.9.1 Intervening event [F(a)1]

An intervening event could create a 'pause' which should, if the circumstances warrant it, be taken into account and excluded when determining the period over which interest is be awarded. It might generate entries such as:

> [F(a)1]
>
> [*no.*] I have found as a fact that during the process of negotiations on this issue [these issues] the claimant stopped that process entirely between [*date*] and [*date*] whilst it was preoccupied in reorganising its accounting systems. In consequence I have determined that the said interruption lay at the claimant's door and constituted an intervening event so as to cause a gap in the period for which allowable interest runs. It is appropriate, therefore, that interest should exclude the effect of that period of interruption.

and the equivalent of paragraph 43 in the Chapter 5 illustration could then be:

> **43** **In the exercise of my discretion, therefore, I shall award on the sum of £**[amount] **interest** [state whether simple or compound, and if compound with what rests] **at** [rate] **per annum, from** [date] **until the date of this my award, but excluding the period from** [date] **to** [date]**, and I calculate interest as £**[amount].

Note in the first of the above two paragraphs the use of 'exclude the effect of'. If the time and/or length of the interruption were such that at the time when the right to interest recommenced, the rate was markedly different from that which would have applied had there been no such interruption, it is likely that one or other party would have made an application that the applicable rate should be that which would have applied during the time of the interruption. The arbitrator would have to hear submissions on that point and decide accordingly (and paragraph 43 would be expanded to reflect that).

Section G

9.3.10   Costs   [G]

9.3.10.1   Award on reserved liability for costs/rejected offer
of settlement   [G(k)]

The effect of a rejected offer of settlement of the substantive claim was briefly discussed in Chapter 4 at 4.3.4.1(4), and awards on reserved matters in general at 7.5. Assume for present purposes that there had been, unknown to the arbitrator, a valid offer to settle made by the respondent and that such offer had been rejected (or not accepted). The arbitrator had been asked to publish an award on the substantive issues but reserving costs, and had published such an award. Assume that his award was less than the rejected offer, but the arbitrator would not at that time be aware of that fact. He had then received submissions from the parties on liability for costs, the submission from the respondent notifying the arbitrator of the rejected offer and seeking an appropriate award on costs, on the basis that as the claimant would have been as well off (in fact better off) had it accepted the offer, it should not be awarded costs from the date by which the offer could reasonably have been accepted. Furthermore, that the claimant should pay the respondent's costs from that date. The claimant simply demurred. In both instances the usual qualifications and exceptions regarding costs would have to be considered. If the arbitrator decided that the allocation was to be as contended by the respondent, and having set down basic recitals in the manner described at 7.4.2, the relevant paragraphs here could be, for example:

[G]

[*no.*] Having been requested to issue an award on all substantive issues but reserving costs, my relevant Award No [*number*] was taken up by the parties on [*date*]. I have since had submissions on liability for costs and have been notified of a valid offer of settlement made by the Respondent. That offer was not accepted within a reasonable time or at all. It dealt properly with interest and costs. The substantive sum plus interest in the offer was greater than the amount I awarded plus interest. There were no other factors which could influence liability for costs.

[*no.*] **I shall therefore award that:**

the Respondent shall pay the Claimant's costs incurred up to the date by which the offer could reasonably have been accepted (which I find to be [*date*]**); and**

the Claimant shall pay the Respondent's costs incurred thereafter.

The operative section H of the award would follow in the normal way, but dealing only with this reserved matter of liability for recoverable costs. After 'I AWARD AND DIRECT THAT' the operative section could repeat the wording of the two indented paragraphs shown above, additionally stating the basis (see 5.3.11.7 at (d)), and conclude with the reservation of determination of the amount (and post-award interest) failing agreement.

Section H

9.3.11   *Operative section*   [H]

9.3.11.1   Time for performance/compliance   [H(d)]
Some arbitrators use the term 'forthwith', intending it as meaning 'without delay'. Others prefer to be specific, stating a period running from the date of the award or from the date when either party has taken up the award, or within some specified period (see illustrative paragraph 48 at 5.3.11.4), thus giving a specific time for compliance. If, for example if holding security for his fees, the arbitrator is in a position to, or for other reason does, issue his award as soon as it is ready, he will be in a position, if he considers it appropriate, to fix a specific date for compliance with the award.

9.3.11.2   Post-award interest – starting date   [H(e)]
If, whilst the award gives a period for payment, it is not intended to or is considered inequitable to allow such a 'break' in the time during

which interest runs, post-award interest can be set to run from the date of the award even if a later date is allowed for payment.

9.3.11.3   Release or disposal of security for costs   [H(g)]

The arbitrator might under section 38(3) have ordered a party to provide security for the other party's costs of the arbitration. In such a case, the form which such security took might not be known to the arbitrator. It might, for instance, be by cash on deposit, or some form of guarantee or legal charge.

If the party which had provided that security is then not found liable for costs, the award must ensure timely release of the security in whatever form it takes. The same applies, effectively to the balance, if that party is found liable for costs, but those costs, when known, are less than the amount of security. If on the other hand those costs, when ascertained, exceed the security, that paying party must make up the balance (although the actual process will depend upon the nature of the security).

The essential point is that the action required must either be included in the award dealing with costs, or be specifically reserved to a subsequent award following submissions on the point.

9.4   Tribunals other than a sole arbitrator

(A brief general note)

As indicated in Chapter 1, because many, possibly all, arbitrations likely to be encountered by the average reader will be before a sole arbitrator, the singular term has been used in this book. Additionally, the narrative and illustrations are all related to a sole arbitrator.

However, in some areas of activity and in some jurisdictions, tribunals of more than one arbitrator are common. Even a cursory perusal of the Arbitration Act 1996 will show a reader that more is involved than simply using the term 'tribunal' instead of 'arbitrator', or 'we' instead of 'I'. Appointments can be made in a variety of ways, often by each party appointing an arbitrator to the tribunal, but other factors impinge upon the award, not least the decision-making process.

To seek to embrace even the basic differences would be to make this book too cumbersome, and could confuse a newcomer to the topic. However, in these increasingly global days, any potential arbitrator should in this context of awards be at least aware of the relevant 1996 Act provisions relating to appointment, decision-making and the like, for tribunals other than a sole arbitrator. These are principally in, but not limited to, sections 15–29 under the heading of 'The arbitral tribunal' (noting sections 20(3) and (4), 21(4) and (5) and section 22, on decision-making), section 52(3) (signing), section 52(5) (seat), section 53 (where the award is treated as made, which is of importance in international arbitrations), and section 54(2) (date of signing).

9.5 **Multi-party arbitrations** (A brief general note)

So as to avoid over-complication, the basis of most of the contents of this book has been an arbitration between two parties. That is not always the case, even though many arbitrators will never encounter anything else. In some instances there can be three or more parties, sometimes, if but rarely, a considerable number. These possibilities can occur in various permutations, with varying effect upon awards. The following are indicative of just some of the possibilities:

(a)   A single claimant against two (or more) respondents:

  (i)   where, if proved, only one of the respondents can be found liable; *or*

  (ii) where, if proved and if the arbitrator is so authorised, liability can be apportioned in different ratios between the respondents.

(b)   Two or more claimants against a single respondent:

  (i)   where, if proved, the respondent can be found liable to only one of the claimants; *or*

  (ii) where, if proved and if the arbitrator is so authorised, the respondent can be found liable in different ratios to the various claimants.

(c)   Two or more claimants against two or more respondents, with a similar but wider permutation of possibilities.

(d)   The various parties are in a 'string', perhaps with each one claiming through the next, with, for example, Party A claiming against Party B who in turn is claiming (possibly with further issues or amounts added) against Party C, and so on. (See also at 9.3.4.2.)

(e)   In each situation, there might or might not be one or more counterclaims.

The complications in such cases might sometimes be substantially procedural rather than relating to decision-making and the award(s). In some instances, however, the award(s) can have substantial difficulties and pitfalls, demanding the exercise of considerable skill and perception by the arbitrator.

Depending upon the nature of the appointment, submissions and issues, there might be a single composite award, or in some instances a series of awards which collectively resolve the totality of the matters in issue. The elements of recital and decision-making discussed in earlier chapters apply equally in these circumstances, but adapted and marshalled to suit the particular circumstances. Care is needed to avoid overlap, duplication, or gaps, and the precise identification of the various, possibly interlinked, issues becomes of considerable importance.

The purpose of this brief mention is simply to put the less experienced readers on notice of the possible situations, so that in any such arbitration the future award is kept in mind when at an earlier stage planning the nature and sequencing of pleadings and submissions, and when later dealing with the recording and collation of information for decision-making.

9.6 Effect of party autonomy

In Chapter 1, it was emphasised in the last paragraph of 1.2 that, as a consequence of the substantial party autonomy provided by the 1996 Act, powers and procedures – and therefore the nature and content of awards – can vary from one arbitration to another. To give cohesion within this book, however, the narrative and illustrations are based almost entirely, with a few identified exceptions, on the 'default' provisions of the Act.

The possible intermixing of default and party-agreed provisions can give innumerable permutations, thus creating differences from some of the comments and illustrations in these pages. That situation is a supplementary purpose of the cross-referencing of the narrative in this book to the related sections of the Act. The cross-referencing should enable any reader to apply the general principles and illustrations set out here, but suitably adapted to the specific provisions of a particular arbitration and any party agreements on procedures or other matters embraced within it.

Part F
Appendices

Appendix 1
Expanded checklist from Chapter 3

The illustration of a possible expanded checklist in Chapter 3 is interspersed with intervening narrative. For convenience it is repeated here as an entity. Any checklist is but a start to the task of deciding the most suitable format and sequence for an award. As stated in the main text, those matters can vary, not only to suit individual arbitrators' preferences, but also to suit the nature of the particular dispute, the nature of submissions and above all the needs of the parties.

First, a possible broad structure:

A   Heading;
B   Background:
     B1   Identification and jurisdiction;
     B2   Interlocutory procedural matters;
     B3   Hearing;
C   Submissions and evidence/result of any inspection(s);
D   Analysis/findings/reasons/decisions, *issue by issue*;
E   Value Added Tax implications;
F   Interest;
G   Costs;
H   'Operative' section (i.e. that giving effect to the award);
I   Reserved matters;
J   Signature and formalities.

Then, as an expanded version:

A possible basic overall sequence and expanded checklist

Heading   [A]

A

(a)   the applicable Arbitration Act;
(b)   any other relevant statute, or scheme of arbitration, or rules;
(c)   identification of the parties.

Background   [B]

Identification and jurisdiction   [B1]
B1

(a)   identification of the parties;
(b)   identification of the contract, or other relationship between the
       parties;
(c)   the law governing the contract or other relationship;
(d)   the essential provisions of the arbitration agreement/arbitration
       clause;
(e)   the provisions for appointment of an (the) arbitrator(s) (including
       any special qualifications of, or requirements relating to, appoint-
       ees – and any incorporated agreement on powers and proced-
       ures);
(f)   any rules incorporated from the outset (section 4(3));
(g)   the fact of a dispute having arisen; any known counterclaim;
(h)   the *general* nature of the dispute(s) and the matters in issue;
(i)   the resultant appointment – and any necessary details and terms;
(j)   any challenge to jurisdiction and how dealt with (sections 30–32);
(k)   basis for deciding the dispute (section 46);
(l)   other matters to the extent they are relevant to this sub-section.

Interlocutory procedural matters   [B2]
B2

(a)   reminder to transfer appropriate items (if any) to a later section;
(b)   the 'seat' of the arbitration (sections 3 and 52);
(c)   any interlocutory agreement(s) regarding adoption of rules,
       or on powers or procedures (sections 4(3), 34–41), e.g. (but not
       restricted to):
       1   on procedural and evidential matters (section 34) (insert
            detailed sub-list if appropriate);
       2   on consolidation or concurrent hearings (section 35);
       3   on appointment of experts, legal advisors or assessors (section
            37);

4   relating to property the subject of the proceedings (section 38(4));

5   on preservation of evidence (section 38(6));

(d)   if appropriate, which 'uncommon' powers have been used, and how;

(e)   any necessary clarification or amendment of the matters in issue;

(f)   any previous award(s) and what issues were decided by each (section 47);

(g)   the issue(s) (claim and any counterclaim) to be decided by this award;

(h)   meetings: preliminary, further, pre-hearing review;

(i)   consequent and subsequent applications and orders re (e.g.):

1   attended hearing, or 'documents only' (section 34(2)(h));

2   type of statements to be exchanged (section 34(2)(c)); timetable;

3   disclosure and production of documents (section 34(2)(d));

4   exchange of witness proofs and experts' reports (section 34(2)(f));

5   meeting(s) of experts (section 34(2)(f));

6   if appropriate, language and translations (section 34(2)(b));

7   questions to be put, when and in what form (section 34(2)(e));

8   other relevant matters;

(j)   resultant party action:

1   brief details of exchange (section 34(2)(c));

2   admissions or agreements; result of meeting(s) of experts;

3   any resultant reduction in, or expansion of, the matters in issue;

(k)   other directions and/or administrative actions if and to the extent not already outlined and if relevant to the award, e.g:

1   orders for security (section 38(3) and 41(6));

2   limitation of costs (section 65);

3   any valid 'costs in any event' agreement (section 60);

4   any 'costs in any event' directions (section 61(2) exception);

(l)   any provisional orders issued under section 39 (if empowered);

(m)   any party default (and, here or later, the consequence), e.g:

1   inordinate or inexcusable delay (section 41(3));

2   failure to comply with any order or direction (section 41(5));

(n)   issue of peremptory order(s), compliance or default (section 41(5) and (6)), consequence (section 41(6) and (7));

(if appropriate, e.g. dismissing claim, repeated in operative section H); court enforcement (section 42);

(o)   other matters to the extent that they are relevant to this subsection.

Hearing   [B3]

B3

(a) reminder to transfer appropriate items (if any) to a later section;
(b) date and place of the hearing (or if dealt with on the basis of documents and written submissions only, or on some other basis – the arrangements);
(c) failure to attend (or equivalent)/any ex-parte proceedings (section 41(4));
(d) who represented the parties (section 36); witness names and designations;
(e) whether oral evidence was taken on oath or affirmation (section 38(5));
(f) if not recited in B2, any default, and any orders issued, during the hearing;
(g) inspection(s) etc. (and who present) (section 38(4));
(h) any agreed (and final) list of issues; any payments during the reference;
(i) any other matters to the extent that they are relevant to this sub-section.

Submissions and evidence   [C]

C

(a) the contentions of the parties (claim and any counterclaim);
(b) a brief but adequate précis or summary of the respective advocates' opening and closing submissions, whether in writing, oral, or a combination;
(c) if not, or to the extent not, covered in Section B:
   1 identification of each witness, with status, qualifications, etc;
   2 brief précis of *relevant* evidence (including any admissions under cross-examination);
   3 equivalent description in ex-parte proceedings;
(d) if not, or to the extent not, covered in Section B, reference to whether the arbitrator had made any inspection of the subject matter or the place concerned ('the locus'):
   1 the purpose of the inspection and who present;
   2 if other than simply to put the matter into context, the result of inspection;
(e) any other matters to the extent that they are relevant.

Analysis/findings/reasons/decisions on the substantive issues   [D]

       D

(a) a numbered list of the issues dealt with in this award (unless sufficiently detailed in the recitals);

(b) any common ground or undisputed facts;

(c) an adequate, but brief, précis of and analysis of the submissions and evidence (unless it is to be an award without reasons), effectively thereby determining (in respect of liability and, if then relevant, the financial or other consequences) and giving the reasons for:

1 findings of fact;

2 what held as matters of law;

3 application of the law to the facts;

4 consideration of any interest claimed as special damages;

5 the decision on the substantive issue(s), also showing why any contentions were rejected;

6 consideration of whether a counterclaim is a set-off; (unless considered under Section G (Costs));

(d) a summary of those decisions (if there are several to draw together);

(e) where money is awarded, effect of any amounts claimed but paid during period of arbitration.

Value added tax implications   [E]

       E

(a) check whether there are VAT implications and, if not dealt with in 'pleadings', request parties' submissions;

(b) seek party agreement on *manner* of treatment, including whether to be in present award, or to be reserved for later submissions;

(c) adopt appropriate procedure (see Chapter 4, Section E).

Interest   [F]

       F

(a) consideration (on claim and counterclaim) of any allowable interest to be added, the rate applicable and whether it is to be simple or compound:

1 on any amount awarded (section 49(3)(a), or contractual, or at common law);

2 on any amount claimed in, outstanding at the commencement of, and paid during the period of, the arbitration (section 49(3)(b)) – and see D(e) above;

      3 on the outstanding amount of any award (i.e. post-award interest – section 49(4)); (from the date of the award, or later date fixed, until payment – on outstanding amount of the award, on any interest awarded under section 49(3), and on any award as to costs);

(b) decision(s), computations of amounts, and summary if appropriate.

Costs [G]

G

(a) consideration of liability for recoverable costs (arbitrator's fees and expenses, fees and charges of any arbitral institution concerned, legal or other costs of the parties) (sections 59–64);

(b) consideration of validity of any agreement to pay costs in any event (section 60);

(c) consideration of whether more than one 'event' (section 61(2));

(d) if not dealt with earlier, consideration of whether a counterclaim constitutes a set-off (affects 'event': section 61(2));

(e) principle adopted and reasons if not following the event (section 61(2));

(f) basis of determination of recoverable costs (section 63(5));

(g) effect of any 'costs in any event' orders or agreement;

(h) effect of any limit on recoverable costs (section 65);

(i) effect, and disposition, of any security provided (section 38(3));

(j) effect of any earlier provisional order(s) for payment of costs (section 39);

(k) effect of any rejection of offer in settlement of substantive dispute;

(l) decision(s) on liability for costs – claim and counterclaim (section 61);

(m) determination (if appropriate at this stage) of *amount* of recoverable costs (section 63(3) – and, if not determined in principle at F(a)3 above, interest on costs (section 49(4)).

Operative section (*commonly headed Award*) [H]

H

(a) the fact of having considered submissions;

(b) 1 if a monetary award, amount (sum plus any interest) to be paid (section 48(4));

      2 if a non-monetary award, the appropriate declaration, direction or, for example, rectification (sections 48(3) and (5));

(c) any conditions or terms;

(d)   time for performance;

(e)   provisions relating to post-award interest (section 49(4)) – and see F(a)3 and G(m) above;

(f)   VAT implications if appropriate;

(g)   who is to pay what costs (section 61) (and, if a party has already paid any fees of the arbitrator(s) for which it is not liable, provision for reimbursement together with any interest relating to such reimbursement).

Reserved matters [I]

I

(a)   Matters left over to a subsequent award.

Signature and formalities [J]

J

(a)   Date when award was made (and, if different, date when signed); seat, if not previously designated (sections 52(5), 53 and 54);

(b)   Signature (section 52(3));

(c)   Witnessed.

Appendix 2 (referred to in Chapter 4)
Illustration of manual collation of location of submissions and evidence

It is unlikely that an IT-generated collation will be in a similar form. It might be based upon key words.

Index of issues: Submissions and evidence: *Techsolut v Sophisticated Theatres*				
Issue: *1(a) Was the work included in letter of enquiry?*				
CLAIMANT				
Submns/Advocate	Documents	Witnesses		
St of Cse: para 8,12	*Bdle A2: 143*	*GJ Wyttle*		
Opng: para 10,24	*W Stat: GJW 27*	Ch	*1: 11–16*	Ch
Clsg: para 17		X-ex	1: 19–32	X-ex
Oral sub: Bk 1,17		Re-x	2: 3–5	Re-x
		Ch		Ch
		X-ex		X-ex
		Re-x		Re-x

[Repeated here for respondent – or more space for both]				
RESPONDENT				
Submns/Advocate	Documents	Witnesses		
St of Df: para 7,15	*Bdle A2: 143*	*H Trimble*		
Opng: para 14–20	*W Stat: HT 14*	Ch	*2: 9–14*	Ch
Clsg: para 19		X-ex	*2: 15–20*	X-ex
Oral sub: Bk 2,32		Re-x	*None*	Re-x
		Ch		Ch
		X-ex		X-ex
		Re-x		Re-x

Explanation of references

The references in the submissions ('Submns') section here refer to the statement of case (or defence), written opening and closing addresses, and advocates' oral submissions; the numbers refer to the document pages, then (bottom entry) the relevant evidence book and page number.

The references in the documents section indicate bundle number and page in that bundle, the initials of the witness and the paragraph of his/her written statement.

The references in the witnesses section refer to examination-in-chief, cross-examination, and re-examination; the numbers refer to the relevant evidence book and page number(s).

Appendix 3 (referred to in Chapter 6)
Substantive monetary award from Chapter 5

The award synthesised in Chapter 5 is fragmented by intervening broader explanatory narrative. It is repeated here as a homogeneous entity, broken only by section headings. Those sections are solely for ease of cross-reference between different chapters.

It is important to note that several of these illustrative award paragraphs in Chapter 5 were subject to caveats or qualifications. The version in this Appendix should not be adopted for guidance without reference to those caveats and qualifications, particularly but not restricted to those relating to paragraphs 6, 40, 48(c), (d) and (e).

In this version a simple contents list has been inserted. In a longer or more complex award such a list can to advantage be more detailed, whether or not further sub-headings are used in the award itself.

This list of contents can either immediately follow the cover page (here omitted) or be added at the end, preferably the former.

Contents

[Section A]

In the matter of the Arbitration Act 1996

and in the matter of an Arbitration

between

Techsolut (GB) plc   **Claimant**

and

Sophisticated Theatres Ltd   **Respondent**

FIRST AWARD

Preamble

The dispute refers to work executed by the Claimant, a company specialising in the design and construction of theatre equipment, at Theatre STL at Weiringhampton on behalf of the Respondent company, which owns and operates a number of specialist theatres.

The Claimant contracted to design, construct and install two complete 'hi-tech' mobile stage sets, including what were termed 'integral finishes'. In addition, they redecorated the permanent stage structure and related backstage areas, then sought to charge extra for that work, alleging that it had been ordered as an extra. The Respondent contended that the redecoration was, or should have been, included in the contract price and also contended that the charge for that work was too high.

Being unable to compromise their differences, they have submitted the dispute to be resolved by arbitration.

[Section B1]

Identification and jurisdiction

1   The Claimant, Techsolut (GB) plc [of *address*], is a company specialising in the design and construction of theatre equipment; the Respondent, Sophisticated Theatres Ltd, [of *address*] owns and operates a number of specialist theatres.

2   The Respondent engaged the Claimant to design and construct two hi-tech stage sets for use in and subsequent to the high profile reopening after refurbishment (by others) of Theatre STL at Weiringhampton. The design, construction and installation of the sets were the subject of an agreement in writing, dated 5th November [*year*], evidenced by a letter of enquiry dated 1st October [*year*], a brief quotation dated 10th October [*year*], and a letter of acceptance dated 5th November [*year*].

3   The contract included an arbitration clause whereby in the event of any dispute arising under the contract such dispute was to be submitted to arbitration by an

arbitrator to be agreed or, if not agreed within [*days*], to be appointed on the application of either party by the President or a Vice-President of [*name of body*] {The Theatre Industry Contracts Tribunal (TICT)}.

4   Following correspondence between the parties' representatives the parties agreed, by letters of [*date*] and [*date*], that any such arbitration should *not* be subject to the Arbitration Rules issued by the TICT.

5   A dispute has arisen within the parameters of the arbitration agreement, its nature and matters in issue being as follows:

On completion of the contract works it was alleged by the Claimant that during the progress of the works of installing the sets:

(a)   extra works were ordered (being backstage redecoration of surfaces not forming parts of the sets for which it was responsible and not being the type of work it normally undertook), without fixing any price or basis of pricing; and

(b)   this caused difficulties for the Claimant, not least because the extra work was outside its province and expertise, even though its staff were multi-skilled, specialist technicians; and

(c)   it valued the said extra work at £18,750 plus VAT, which amount it sought to recover from the Respondent; and

(d)   the Respondent has refused to pay that or any sum in respect of the said extra work; and

(e)   the Claimant now claims:

  (i)   that sum of £18,750;

  (ii)   Value Added Tax related to that sum;

  (iii) interest under section 49 of the Arbitration Act 1996; and

  (iv)   costs

and the Respondent contends that:

(f)   the said work was, or should have been, included in the Claimant's contract figure, which has been invoiced and paid;

(g)   it did not at any time order the said work as an extra;

(h)   even if, which it denies, it is found liable to pay for the said work the amount claimed is excessive; and

(i)   the Respondent claims costs.

These matters are the subject of this arbitration.

6   Following a joint invitation signed by the parties and dated 2nd April [*year*] I, Jonathan Fairley MSc (Performing Arts), of Palladium House, Laburnum Way, Nevekirk, forwarded to both parties a form of appointment which, inter alia, set out my terms and conditions. That form was returned to me, signed by the parties on 9th and 10th April [*year*] respectively – and I signed it by way of acceptance of appointment on 12th April [*year*], thereby taking up jurisdiction in this reference.

[Section B2]

Interlocutory procedural matters

7   The parties have, by written agreement, designated the Seat of the arbitration as being England {and Wales}.

8   The issue to be determined by my award is: *Is the Claimant entitled to additional payment for executing the work in dispute and if so how much?* The parties have agreed that the sub-issues to be resolved in determining that issue are as follows and that I am to award such amount, if any, as I so determine:

(a)   was the disputed work included in the Respondent's letter of enquiry dated 1st October [*year*]?

(b)   if not, was it specifically included in the Claimant's quotation dated 10th October [*year*]?

(c)   if neither, did the Respondent subsequently order that work {it having been agreed that no issue is taken as to whether or not the Respondent was entitled to issue such order}?

(d)   if so, did the said work constitute either extra work or a collateral or other contract (but for this purpose without any need to decide which) entitling the Claimant to extra payment?

(e)   if so, what amount was a fair charge for that work?

9   I conducted an initial preliminary meeting on 19th April [*year*] at my office at which the parties were represented by:

Claimant   :   Mr G.J. Wyttle, Technical Director;

Respondent :   Mr Justin Blackwelle, Counsel,

following which I issued my Order No 1 giving directions for the conduct of the reference, providing for:

> an attended Hearing;
> exchange of Statements of Case;
> limited disclosure and production of documents;
> exchange of witness proofs and experts' reports; and
> meetings of experts.

I issued further orders from time to time, two of them with a sanction as to costs (being Orders Nos [*number*] and [*number*]).

10   Statements of case were subsequently exchanged, as were witness statements, experts' reports and such documents as had been requested or ordered to be produced. The consequence of delay in production is referred to at paragraph 11 below. In the course of these exchanges and relating to Value Added Tax, the Claimant amended that part of its claim to a request for a declaratory award.

11   My Orders Nos [*number*] (failure to produce documents timeously) and [*number*] (abortive application) provided that the Respondent be liable for the costs of and resulting from those orders in any event.

12   The parties agreed that in view of the relatively modest amount in dispute there was no need for a pre-hearing review, but I was notified by a joint letter dated 20th May [*year*] that they had agreed to limit the length of the Hearing to one-and-a-half six-hour days plus a site inspection; the time to be shared equally between them, but 'with a reasonable amount of give and take and with any differences as to the sharing of time being resolved by the arbitrator on an instant but fair basis'.

[Section B3]

Hearing

13   The hearing was held on Tuesday and Wednesday 24th and 25th July [*year*] in the Chamber of Commerce Council Chamber at Weiringhampton.

14   Both parties were represented by Counsel, Ms Susan Wellbie for the Claimant and Mr Justin Blackwelle for the Respondent. Opening addresses had been submitted in advance by both Counsel, and closing submissions were also submitted in writing in sequence.
     The various witnesses were:

     Called by the Claimant:

          Witnesses of fact:

               Giles Jay Wyttle   :   Technical Director
               George Amble   :   Multi-skilled technician

          Expert witness(es):

               Arthur Worsnip   :   Hi-tech set specialist

     and by the Respondent:

          Witnesses of fact:

               Hyram Trimble   :   Managing Director
               James Bull   :   Stage Manager

          Expert witness(es):

               Jon Chekov   :   Independent producer

Witness statements and experts' reports having been exchanged, they were, as had been agreed, taken as read as evidence-in-chief, but with oral confirmation and limited expansion.

15   Oral evidence was generally taken on oath. Instances where a witness requested instead to affirm are mentioned in the relevant section of this award only if and where necessary to explain my reasoning or conclusion.

16   On completion of the hearing on 25th July [*year*] I visited Theatre STL, accompanied by Mr Worsnip and Mr Chekov, and inspected the work of decoration in question and the sets. My attention was drawn to:

(a)   the nature of the mobile sets and their integral finishes;

(b)   the extent and degree of complexity of the decoration to all spaces behind the line of the proscenium opening (including stage equipment and gantries); and

(c)   the nature of access to those areas;

the purpose of my inspection being solely to familiarise myself with these matters, not to make decisions on the basis of that inspection.

[Section C]

Submissions and evidence

17   **Relating to sub-issues (a) and (b):**

(Was the disputed work included in the Respondent's letter of enquiry dated 1st October [*year*]? If not, was it specifically included in the Claimant's quotation dated 10th October [*year*]?):

The Claimant contends that:

(a)   the backstage redecoration, not forming part of the sets as such, was not referred to in the letter of enquiry upon which the Claimant's tender was based, nor was it included in that tender; further that such exclusion was apparent on the face of the tender;

(b)   the Respondent was at all relevant times aware that the said works of decoration constituted work of a nature not normally executed by the Claimant;

(c)   whilst pre-invitation there were wide-ranging discussions between the parties concerning the whole of the stage area and its equipment, only the mobile sets were mentioned in the letter of enquiry;

(d)   the said works of redecoration were not a necessity for the proper functioning of the mobile sets.

The Respondent contends that:

(a) it was at all times clearly understood and acknowledged by the Claimant that the Respondent required the backstage areas to be decorated at the same time as the installation of the sets, and there was never any question of that work being done by others;

(b) there was no specific exclusion of that work in the Claimant's quotation dated 10th October [*year*];

(c) there was a necessary inference to be drawn that the said works of redecoration were an essential prerequisite for the functioning of the mobile sets.

18 **Relating to sub-issues (c), (d) and (e):**

(If neither, did the Respondent subsequently order that work? If so, did the said work constitute either extra work or a collateral or other contract entitling the Claimant to extra payment? If so, what amount was a fair charge for that work?)

The Claimant contends that:

(a) the Respondent's agent, Mr Hyram Trimble, who had signed the letter of acceptance of 5th November [*year*], gave clear oral instructions for the execution of that work and when so doing was aware that the Claimant's schedule of work included in the accepted quotation made no reference to that work;

(b) it is entitled to extra payment for the said work;

(c) the amount claimed represents a fair quantum meruit evaluation.

The Respondent contends that:

(a) whilst acknowledging that Mr Trimble was its accredited agent, his comments to Mr Amble had constituted nothing more than clarification, not intended to be an order for extra work;

(b) in consequence no extra payment is due therefor;

(c) if contention (b) is rejected, a fair figure is little over £10,000.

[Section D]

My consideration of the issues

The sole issue: *Is the Claimant entitled to additional payment for executing the work in dispute and if so how much?*

Sub-issues (a) and (b): *Was the disputed work included in the Respondent's letter of enquiry dated 1st October* [*year*]*? If not, was it specifically included in the Claimant's quotation dated 10th October* [*year*]*?*

19   It is convenient to take these two sub-issues together.

20   It is common ground that:

(a)   the decoration to backstage surfaces was executed by the Claimant;

(b)   a colour schedule was passed to the Claimant by the Respondent; and it is not in issue that:

(c)   the letter of enquiry as such made no specific mention of decoration to backstage surfaces.

21   The Respondent's Managing Director, Mr Trimble, contended that he had repeatedly during earlier discussions mentioned the required visual impact between the mobile sets and the blank rear wall. He was vague as to just when those discussions took place.

22   To the extent that, if at all, such discussions were relevant, I accept as the more convincing Mr Wyttle's evidence (and I find as facts) that those discussions were not only of a general nature and preceded any invitation to submit a quotation, but also related to the visual link between the proscenium opening, the curtains (which were never any part of the work to be quoted for by the Claimant), the sets in their various configurations, and the rear (and any other visible) walls.

23   Mr Blackwelle, for the Respondent, submitted that the contract was partly oral ('a meeting of minds in those discussions') and partly written ('the enquiry, the quotation – i.e. the offer – and the acceptance'). Ms Wellbie, for the Claimant, argued that the parties by those written documents had put their agreement into precise terms.

24   The letter of enquiry stated 'Following our various meetings, I am setting out the parameters to our requirements: a dimensioned sketch design to be developed by you; functions; rotates and other movables required; minimum needs re power outlets; computerised facility requirements; and colours of exposed – i.e. visible in any of the configurations – parts.'

25   The quotation incorporated that letter by reference, added technical specifications and outline working drawings; it made no reference to backstage redecoration, or to discussions or to other agreements. Those matters are not in contention.

26   It is clear that the enquiry and quotation did put the agreement in precise terms. The enquiry was in sufficient detail as to performance and extent as to make it inconceivable that other work of any consequence could have been intended but not mentioned. The works of redecoration (much of which was at all times, even on an open stage, not visible from the auditorium) were clearly not a prerequisite for the functioning of the mobile sets, there was no inference to be drawn, **and I so hold**. It is therefore inappropriate for me to consider oral evidence of some alleged different or additional intent, but even if that were not so, I am satisfied from Mr Wyttle's evidence that there was no such intent by either party.

27   In consequence, I hold that the letter of enquiry must stand by itself and be so interpreted.

28   *I determine that the disputed work was not included in the said letter of enquiry, nor was it included in the Claimant's quotation, whether specifically, by reference, or by necessary inference.*

Sub-issue (c): If neither, did the Respondent subsequently order that work?

29   Mr Amble, the Claimant's technician who supervised the site assembly of the sets, gave evidence that on many occasions Mr Trimble said things like *'I want you to paint these walls and those others before you finish assembly of the sets'*, to which he had replied to the effect of *'I know nothing of that'* in response to which Mr Bull, the Stage Manager, had been asked by Mr Trimble, in his presence, to give him (Mr Amble) a colour schedule. That colour schedule (document C27) is headed *'Paint the surfaces to the following colours:'* and was signed by Mr Trimble. I found Mr Amble's evidence convincing, but that of Mr Trimble and Mr Bull vague and evasive on this point. Mr Amble said that he took the repeated requests by Mr Trimble and the reaction to his own response, including the issue of that schedule, as constituting an order to do the work. He also told us that he had then said *'OK. I'll ask the boss to raise a works number.'*, to which, he says, he got the response *'Yes, do that'*. Once again I find Mr Amble's evidence convincing.

30   In consequence, I hold that Mr Trimble's words and actions led Mr Amble, quite properly, to believe that extra work had been ordered. Neither Mr Trimble nor Mr Bull did anything to disillusion him in that belief. In those circumstances any reasonable person would justifiably have taken it that he had received an oral order.

31   *I determine that the Respondent did subsequently order that work.*

Sub-issue (d): If so, did the said work constitute either extra work or a collateral or other contract?

32   It was made clear to me by both Counsel that the term *'extra work'* was intended simply to refer to work ordered as a variation of or under the contract as contrasted with a separate contract of some kind. They agreed that nothing now hangs on the distinction, and that, if I find the work to be extra work, I am not requested to determine its precise contractual status.

33   I have no hesitation in concluding that the work was 'extra' to the extent that, as I have found, it was additional to the contract work, it was ordered by the Respondent – and that the Claimant is entitled to extra payment for it.

34   *I determine that the said work did constitute extra work of a nature in respect of which the Claimant is entitled to be paid.*

Sub-issue (e): If so, what amount was a fair charge?

35   Page D43 of the agreed bundle was a copy of part of the Claimant's computation of the amount charged and claimed. I pointed out that there appeared to be
a duplication of the entry for one operative and it was conceded by Ms Wellbie
that that was so. That would have had the effect of reducing the amount claimed
to £18,150.00.

36   Mr Worsnip, the expert called by the Claimant, had considerable experience
not just of 'mobile' set design but of traditional sets and of re-equipment and (of
specific relevance here) of refurbishment of backstage areas. He had analysed
the successful bids relating to such work at six theatres with which he had been
concerned and had broken down the various figures in a manner which gave
broad unit prices and facilitated comparison with the redecoration at Theatre
STL. On that basis he gave as his opinion that the work here in dispute, had it
been the subject of competitive tenders, would have cost approximately
£16,000. His assessment of the impact of the difficulty of access caused by the
presence of the Claimant's set erectors simultaneously executing their contract
work amounted to a further £750.00. He said that if works had perforce to be
executed by non-painters it could be expected to be more costly, perhaps of the
order of 5% to 7½%. On the other hand, Mr Chekov, the expert called by the
Respondent, simply contended that he could have had the work done for no
more than £10,000, but produced no justification for his contentions. In cross-
examination, he admitted that he had little experience of this kind of work.

37   Ms Wellbie asked me to have regard to the compounding effect of the various
unusual factors: an unexpected order for this work, the need to complete by the
originally-intended date, attempts to find and the non-availability of local painters, the remote nature of the location, and the enforced use of technicians who
were not general painters. She contended that this compounding factor fully
accounted for the difference between Mr Worsnip's figures and the amount
claimed. Mr Blackwelle confirmed that the Respondent admitted the points
concerning the non-availability of local painters and that the work had of
necessity to be executed by the Claimant's technicians, but did not accept that
the Claimant's charge was fair. I find that there is weight in Ms Wellbie's
contention, even if not fully to the extent suggested. Having regard to these
various matters, I have determined that a proper amount for this work, excluding
any applicable VAT, is £17,100.00.

38   *I determine that a fair charge for the work in question is
£17,100.00.*

My decision on this issue.

39   **I determine that the Claimant is entitled to additional payment, in
the sum of £17,100.00.**

[Section E]

Value Added Tax

40   There was no dispute between the parties that if I were to award any sum to the Claimant, Value Added Tax would then apply to some part, or all, of the amount awarded. {However, as negotiations were still in hand with HM Customs and Excise as to whether elements of the work (being to a listed building) were in fact subject to that tax, it would be impossible at present to ascertain the amount of the tax.} I have found a sum to be due under the substantive claim and in consequence hold that the Respondent is in addition liable to pay such amount in respect of VAT as is in due course found to be properly payable under the VAT Regulations, together with interest if such payment is delayed.

[Section F]

Interest

41   I now have to consider interest on the amount I have determined as due to the Claimant. Contractual interest is not applicable. It is, therefore, a matter for my discretion under section 49 of the Arbitration Act 1996.

42   The amounts in question were invoiced (although in greater amounts than I have determined) on 18th February [*year*], a few days after completion of the painting in issue. There was contractual provision for payment against invoices to be made within 28 days. The Respondent has contended that I should not award interest, arguing that the Respondent did not pay on time simply because of what he contended to be grossly excessive charges. I have rejected that contention. Additionally, the Claimant's invoiced amount, though high, was not excessive, nor had it been so would that have been good reason for failing to pay a reasonable amount against the invoice. Regarding interest, then, I shall exercise my discretion in favour of the Claimant, the period to run from 28 days after the date of the invoice (and thus from 18th March [*year*]). As to the rate, having seen credible evidence (which was not disputed) that the Claimant's average bank balances over the preceding years were generally 'neutral', with deposits and borrowing being approximately in equilibrium, I shall adopt a rate slightly above the average base rate over the relevant period (i.e. from 18th March [*year*] to the date of this award). The Claimant at the hearing conceded that in all the circumstances, if I were to allow interest this could be on a simple interest basis. I therefore adopt 5% simple interest.

43   **In the exercise of my discretion, therefore, I shall award on the sum of £17,100.00 simple interest at 5% per annum, from 18th March [*year*] until the date of this my award, which interest I compute at £269.38.**

44   **As to post-award interest, again in the exercise of my discretion under section 49, I shall provide for compound interest to be pay-**

able at [*rate*] % per annum compounded with quarterly rests from the date of this award until the date of payment.

[Section G]

Costs

45   I find no reason to depart from the normal rule. The Claimant was successful in its claims with only a modest reduction in quantum and the only 'costs in any event' orders were against the Respondent. In consequence the Claimant is entitled to its recoverable costs. I have been asked to reserve quantum.

46   **I determine, therefore, that liability for costs (including my fees and expenses) shall follow the event.**

[Section H]

AWARD

47   I have read, considered and taken note of the statements of case, the written and oral evidence and the addresses by Counsel.

48   I AWARD AND DIRECT THAT, in full and final settlement of all claims in this arbitration:

(a)   the Respondent shall within [*number*] days after that when either party shall have taken up this award pay to the Claimant:

  (i)   the sum of £17,100.00 (seventeen thousand one hundred pounds); *and*

  (ii)   interest as computed at paragraph 43 above and amounting to £269.38 (two hundred and sixty nine pounds and thirty eight pence); *and*

(b)   the amounts awarded at paragraph 48(a)(i) and (ii) above (or, in the event of part payment by the due date, the outstanding balance thereof) shall be subject to the addition of compound interest at [*rate*] per cent with quarterly rests from the date of this award up to the date of payment; *and*

(c)   the Respondent shall within [*number*] days of receipt by it of an appropriate VAT invoice from the Claimant pay to the Claimant such Value Added Tax, if any, as is properly chargeable under the applicable VAT Regulations on the principal sum awarded at paragraph 48(a)(i) above, {and interest thereon at [*rate, simple or compound, period of rests*] from the last date for receipt thereof until the date of payment by the Respondent}, *and*

(d)   the Respondent shall pay the Claimant's recoverable costs, together with interest thereon at [*rate, simple or compound, period of rests*] from the date of this award until payment, the amount of such costs and interest thereon (failing agreement) to be determined by me {on the basis described at sub-section 63(5)(a) and (b) of the Arbitration Act 1996} by my further award; *and*

(e)   the Respondent shall also pay and bear my fees and expenses in respect of this my First Award, which fees and expenses I determine at £[*amount; figures and words*], plus £[*amount*] Value Added Tax, thus totalling £[*amount; figures and words*], and to the extent that the Claimant has paid any part thereof the Respondent shall forthwith reimburse him with that amount {and interest thereon at [*rate, simple or compound, period of rests*] from the date of the Claimant's payment until the date of such reimbursement}.

49   Fit for Counsel.

[Section I]

50   I RESERVE to my further or final award determination, if not agreed, of the amount of recoverable costs and any interest thereon (for which purpose I shall give my further directions upon the application of either party).

[Section J]

This my First Award made and published by me {in Nevekirk, England} on [*date*].

[*Signed*]..........*Jonathan Fairley*...........
                Jonathan Fairley, Arbitrator

in the presence of: .*Anne Gee*..................   [Signature of witness]
of.. *The Lodge, Weiring Common, Weirshire.*   [address]
.....*Secretary*.......................................   [occupation]

Index

Numbers shown are section numbers

Printed and bound in the UK by
CPI Antony Rowe, Eastbourne

Printed and bound by CPI Group (UK) Ltd, Croydon, CR0 4YY

16/04/2025

14658830-0004